Progress in Nonlinear Differential Equations and Their Applications
Volume 59

Variational Problems in Riemannian Geometry

Bubbles, Scans and Geometric Flows

Paul Baird
Ahmad El Soufi
Ali Fardoun
Rachid Regbaoui
Editors

Springer Basel AG

Editors' addresses:

Paul Baird
Ali Fardoun
Rachid Regbaoui
Département de Mathématiques
Université de Bretagne Occidentale
UFR Science et Techniques
6, avenue Victor Le Gorgeu
B.P. 809
29285 Brest Cedex
France
e-mail: Paul.Baird@univ-brest.fr
Ali.Fardoun@univ-brest.fr
Rachid.Regbaoui@univ-brest.fr

Ahmad El Soufi
Laboratoire de Mathématiques et Physique Théorique
UMR 6083 du CNRS
Université de Tours
Parc de Grandmont
37200 Tours Cedex
France

e-mail: ahmad.elsoufi@univ-tours.fr

2000 Mathematics Subject Classification: 58Exx, 35Jxx, 35Kxx

A CIP catalogue record for this book is available from the Library of Congress, Washington D.C., USA

Bibliographic information published by Die Deutsche Bibliothek
Die Deutsche Bibliothek lists this publication in the Deutsche Nationalbibliografie; detailed bibliographic data is available in the Internet at <http://dnb.ddb.de>.

ISBN 978-3-0348-9640-5 ISBN 978-3-0348-7968-2 (eBook)
DOI 10.1007/978-3-0348-7968-2

Originally published by Birkhäuser Verlag in 2004
Softcover reprint of the hardcover 1st edition 2004

Printed on acid-free paper produced of chlorine-free pulp. TCF ∞

ISBN 978-3-0348-9640-5

9 8 7 6 5 4 3 2 1

www.birkhasuer-science.com

Contents

Preface

This volume has grown from a conference entitled *Harmonic Maps, Minimal Surfaces and Geometric Flows* which was held at the Université de Bretagne Occidentale from July 7th–12th, 2002, in the town of Brest in Brittany, France. We welcomed many distinguished mathematicians from around the world and a dynamic meeting took place, with many fruitful exchanges of ideas.

In order to produce a work that would have lasting value to the mathematical community, the organisers decided to invite a small number of participants to write in-depth articles around a common theme. These articles provide a balance between introductory surveys and ones that present the newest results that lie at the frontiers of research. We thank these mathematicians, all experts in their field, for their contributions.

Such meetings depend on the support of national organisations and the local community and we would like to thank the following: the *Ministère de l'Education Nationale, Ministère des Affaires Etrangères, Centre National de Recherche Scientifique (CNRS), Conseil Régional de Bretagne, Conseil Général du Finistère, Communauté Urbaine de Brest, Université de Bretagne Occidentale (UBO), Faculté des Sciences de l'UBO, Laboratoire de Mathématiques de l'UBO* and the *Département de Mathématiques de l'UBO*. Their support was generous and ensured the success of the meeting.

We would also like to thank the members of the scientific committee for their advice and for their participation in the conception and composition of this volume: *Pierre Bérard, Jean-Pierre Bourguignon, Frédéric Helein, Seiki Nishikawa* and *Franz Pedit.*

A large conference requires an enormous amount of work behind the scenes; we thank *Annick Nicolle*, secretary for the Laboratoire de Mathématiques at the UBO, for her administrative skills which ensured the smooth running of the meeting.

Finally, more than ninety mathematicians participated at the conference and we hope also that this volume will be a tribute to their enthusiasm.

The organisers

Paul Baird
Ahmad El Soufi
Ali Fardoun
Rachid Regbaoui

Progress in Nonlinear Differential Equations
and Their Applications, Vol. 1, ix–xvii

Introduction: Best Maps and Best Metrics

Paul Baird, Ahmad El Soufi, Ali Fardoun and Rachid Regbaoui

1. Introduction

Among the most beautiful problems of mathematics are those which we can visualise, where the solution contains the faith of our intuition, and yet where achieving this solution proves tantalisingly elusive. Such problems demand the introduction of new techniques and ideas in order to advance. The *uniformisation of geometric structures* has provided such an impetus for the development of differential geometry over many years. We refer back to Riemann and the problem of recognising Riemann surfaces and the conformal structures defined upon them; to the influence of physics leading to the notion of *Einstein metric*; to the problem of identifying *geometric structures on* 3*-manifolds* and that of finding a conformal metric of constant scalar curvature – the so-called *Yamabe problem.* Naturally, in looking for "best" geometric structures, one is lead to look for "best" maps as well, for example *harmonic maps* and *p-harmonic maps*, *eigenvalue problems* or more general solutions of elliptic equations. The unifying feature of all these problems is their *variational characterization.* Solutions arise as extremals of an "action" integral, perhaps with constraints.

The articles in this volume are devoted to the study of such problems; some papers provide an overview whereas others present new work and lead the reader to the very latest developments.

2. Best metrics

The basic object is a manifold M^m; unless we specify otherwise, this will be a *smooth connected Hausdorff paracompact manifold without boundary.* By taking a partition of unity, we can endow M^m with a smooth metric g. We will be concerned with metrics that are *positive definite*, i.e., $g(v,v) > 0$ for every non-zero tangent vector v. Our interest is to find relations between the topology of M^m and its geometry. Here intuition plays a role: we can "see" that a sphere is fundamentally different to a torus, both topologically (there are curves on the torus that cannot be shrunk to a point) and geometrically (the standard sphere "bends" in the same way everywhere, whereas the torus does not). To make these intuitions precise, we introduce the notion of *curvature.*

If ∇ denotes the Levi-Civita connection on (M^m, g), then the *Riemannian curvature* is the 3-covariant 1-contravariant tensor field R^M given by

$$R^M(X,Y)Z = \nabla_X \nabla_Y Z - \nabla_Y \nabla_X Z - \nabla_{[X,Y]} Z \quad (X,Y,Z \in C^\infty TM).$$

If X, Y are orthonormal, the *sectional curvature* of the plane spanned by X and Y is the number $g(R^M(X,Y)Y, X)$.

The *Ricci tensor* is the symmetric tensor field Ric^M defined by

$$\mathrm{Ric}^M(X,Y) = \mathrm{Tr}_g g(R^M(X,-)-,Y) = \sum_{i=1}^m g(R^M(X,e_i)e_i,Y),$$

where $\{e_i\}$ is an orthonormal frame. Finally, the *scalar curvature* is the function $S^M = \mathrm{Scal}^M$ defined by

$$\mathrm{Scal}^M = \mathrm{Tr}_g \mathrm{Ric}^M = \sum_{i=1}^m \mathrm{Ric}(e_i, e_i).$$

For a surface, we have the basic identity:

$$R^M(X,Y)Z = K^M\left(g(Y,Z)X - g(X,Z)Y\right),$$

where K^M denotes the Gaussian curvature of M^2 (the same formula holds in arbitrary dimension, when the manifold M has *constant* sectional curvature K^M). Taking traces, $\mathrm{Ric}^M = K^M g$ and $\mathrm{Scal}^M = 2K^M$, so there is essentially just one curvature for a surface, the Gaussian curvature. (Here we are restricting our discussion to *instrinsic* curvature; of course for an immersed manifold we could discuss the *mean curvature*, for example.)

In dimension 2, we have a clear idea of what we mean by "best" metric – one of constant Gaussian curvature. The *Riemann mapping theorem* (or *uniformisation theorem*) asserts that every simply-connected domain of the complex number sphere S^2 with a minimum of two boundary points can be mapped conformally and diffeomorphically onto the interior of the unit disc (see [21] for a proof of this – apparently Riemann gave no rigorous proof!). The consequences of this theorem are far reaching. For example, every compact Riemann surface M^2, is conformally equivalent to E/Γ, where $E = S^2, \mathbf{R}^2, H^2$ is one of the two-dimensional space forms and Γ is a discrete group of isometries acting properly and discontinuously. This gives us a complete picture of a compact Riemann surface and implies that each one can be deformed in its conformal class to one of constant curvature.

The problem of deforming a metric on a surface to one of constant Gaussian curvature has a variational formulation. Indeed, if (M^2, g_0) is a compact Riemann surface without boundary with scalar curvature S_0, then a conformal deformation of the form $g = e^{2u} g_0$ produces a metric with scalar curvature $S = e^{-2u}(-2\Delta_0 u + S_0)$, where Δ_0 is the Laplace-Beltrami operator with respect to the metric g_0. Thus, in order to find a metric of constant curvature, we are required to solve the partial differential equation

$$-2\Delta_0 u + S_0 = e^{2u}.$$

The left-hand side is now the gradient of the functional

$$I(u) = \int_M |\nabla u|_0^2 \mathrm{d}\mu_0 + 2\int_M S_0 u \mathrm{d}\mu_0 ,$$

subject to the constraint $\int_M e^{2u}\mathrm{d}\mu_0 = S_0 \int_M \mathrm{d}\mu_0$, where $\mathrm{d}\mu_0$ denotes the canonical measure and ∇u the gradient of u with respect to the metric g_0 (see [22], [2]). Hamilton gave an elegant way of obtaining a solution from the evolution equation

$$\frac{\partial g}{\partial t} = (r - S)g , \tag{1}$$

where r is the average value of S [10]; we shall discuss such equations in more detail below.

In higher dimensions, it is therefore natural to formulate the following problem: *given a compact Riemannian manifold* (M^m, g_0), *find a metric* $g = u^{4/(n-2)}g_0$ *with constant scalar curvature.* This problem was first posed by Yamabe in 1960 [26] and is now known as the *Yamabe problem.* Yamabe's work had an error noticed by Trudinger [24], who solved the problem in a special case. Since that time there have been contributions from a number of authors. It was solved completely in the compact case by Aubin [1] and Schoen [19]. The delicate aspect of this problem concerns the critical exponent in the Sobolev embedding theorem, as follows.

The equation to solve is now

$$-\Delta_0 u + \frac{n-2}{4(n-1)} S_0 u = \lambda u^{(n+2)/(n-2)}, \tag{2}$$

giving a metric of constant scalar curvature $S = 4(n-1)\lambda/(n-2)$. This has the following variational characterisation.

Let $H_1^2(M)$ be the Sobolev space of functions in L^2 whose first derivatives are also in L^2. Let I be the functional associated to equation (2):

$$I(u) = \int_M |\nabla u|_0^2 \mathrm{d}\mu_0 + \frac{n-2}{4(n-1)} \int_M S_0 u^2 \mathrm{d}\mu_0 .$$

By the Rellich-Kondrakov theorem, the embedding $H_1^2(M) \hookrightarrow L^q(M)$, $q \in (2, \frac{2n}{n-2})$ is compact; which leads to a solution of the subcritical equation

$$-\Delta_0 u + \frac{n-2}{4(n-1)} S_0 u = \lambda_q u^{q-1},$$

subject to the constraint $u \in H_q = \{u \in H^{1,2}(M) : \int_M u^q \mathrm{d}\mu_0 = 1\}$, where $\lambda_q = \inf\{I(u) : u \in H_q\}$. Subtle arguments are required to extend this to the critical exponent $q = 2n/(n-2)$. Essentially, one is required to show that there is a smooth everywhere positive function u for which the Yamabe invariant $\mu_0 = \inf_{u\in\mathcal{H}} I(u)$ is achieved, where $\mathcal{H} = \{u \in H_1^2(M) : \int_M |u|^{2n/(n-2)}\mathrm{d}\mu_0 = 1\}$ – see [12] for an excellent account. Recent developments concerning critical Sobolev exponents are discussed in the article of E. Hebey in this volume [13].

Although a beautiful problem in dimension $m > 2$, the solution to the Yamabe problem does not give us a conclusive description of manifolds. For example the manifolds S^3 and $S^2 \times \mathbf{R}$ both have constant positive scalar curvature. By factoring through discrete groups of isometries we can obtain compact 3-manifolds which are fundamentally different. From the first, we could obtain the Lens spaces, for example, which cannot be obtained from the second (see [20] for an account of such matters).

In order to look for "best" metrics in higher dimensions, it is natural to consider the functional:

$$\mathcal{R}(g) = \int_{M^m} \mathrm{Scal}^{\mathrm{M}} \mathrm{d}\mu(g) \tag{3}$$

defined on a suitable space of metrics. In order to avoid trivial solutions, we need to impose the constraint $\int_{M^m} \mathrm{d}\mu(g) = 1$, i.e., that the volume remain constant. On taking a variation of (3) with respect to the metric g subject to the constraint, we obtain the Euler-Lagrange equations in the following form:

$$\frac{1}{2}\mathrm{Scal}^{\mathrm{M}} g - \mathrm{Ric}^{M} = ag$$

where a is a constant. Solutions therefore occur when $\mathrm{Ric}(g) = \lambda g$, for some function λ. In dimension $m > 2$, the Bianchi identities now imply that λ must be constant and we are therefore led to the equation

$$\mathrm{Ric}(g) = cg$$

with c constant. A metric satisfying this equation is called an *Einstein metric.* Unfortunately in dimension $m = 3$, this condition does not give us enough flexibility to pick out all compact manifolds, since it implies that M^3 must have constant sectional curvature which in turn implies that $M^3 = E/\Gamma$, where $E = S^3, \mathbf{R}^3, H^3$ is one of the three-dimensional space-forms and Γ is a discrete group of isometries acting discretely and properly discontinuously. However, in dimension 4, the construction of Einstein metrics is a rich field of study, with much recent progress.

In dimension 3 the picture is rather delicate and we are guided by the conjecture of Thurston. A *model geometry* is a smooth simply connected manifold E together with a Lie group G of diffeomorphisms of E which act transitively on E with compact point stabilisers, such that G is maximal in the sense that it is not contained in any larger group of diffeomorphisms of E with compact point stabilisers. It is also a requirement that there exists at least one compact quotient, i.e., there exists a subgroup H of G such that E/H is a compact manifold. Two such geometries are considered equivalent if there exists a diffeomorphism between them which intertwines the group actions. Thurston has classified the geometries; up to equivalence there are eight of them and they are given by $S^3, \mathbf{R}^3, H^3, S^2 \times \mathbf{R}, H^2 \times \mathbf{R}, \mathrm{Nil}, \widetilde{\mathrm{SL}_2(\mathbf{R})}, \mathrm{Sol}$. The *geometrization conjecture* of Thurston asserts that any compact 3-manifold is made up of pieces, separated by 2-spheres or 2-tori, each piece covered by one of the eight geometries. The resolution of this conjecture in the affirmative would imply the Poincaré conjecture.

A powerful result, in the spirit of the above ideas, was proved by R. Hamilton [9]. He established that if (M^3, g) is compact with $\mathrm{Ric}(g) > 0$, then the metric g can be smoothly deformed to one of constant sectional curvature. This means that

M^3 is diffeomorphic to S^3/Γ, where Γ is a discrete group of isometries acting properly discontinuously. His method was to allow the metric to evolve using the so-called Ricci flow:

$$\frac{\partial g}{\partial t} = -2\mathrm{Ric}(g) + \frac{2}{n} rg \qquad (n = 3)\,,$$

where $r = \int_M S \mathrm{d}\mu / \int_M \mathrm{d}\mu$ is the average scalar curvature (this in fact restricts to the evolution equation (1), in the case of dimension 2). Since the right-hand side is a semi-elliptic operator in the components of the metric, this equation should be viewed as a kind of *heat equation.* Intuition suggests that in the limit, we will obtain a metric of constant Ricci curvature. As already noted, in dimension 3, this means that the manifold has constant sectional curvature and the result follows. More recently, in order to attack Thurston's geometrisation conjecture, Hamilton has proposed a program involving the *unormalised* Ricci flow:

$$\frac{\partial g}{\partial t} = -2\mathrm{Ric}(g)\,.$$

The article of D. Knopf [15] in this volume, discusses the very latest work on this method and notably the possible resolution of the geometrisation conjecture by G. Perelman.

Geometrically, fixed points (up to diffeomorphism and dilation) of the Ricci flow are interesting objects (see [15]); they are metrics g which satisfy the equation

$$-2\mathrm{Ric}(g) = \mathcal{L}_X g + ag\,,$$

for a constant a and a vector field X. The article of Gastel [8] shows how to obtain examples of solitons by a warped product construction.

3. Best maps

In seeking "best" maps, we should note that they may be defined in terms of a metric and therefore a "best" metric may play a role – i.e., the two problems are interrelated, and indeed we find that their variational characterisation and description as solutions to an elliptic partial differential equation have many analogies.

Let us briefly recall the main theorem of Hodge theory (see, for example [25]). Let (M^m, g) be a compact Riemannian manifold; then each element of the kth order de Rham cohomology group $H^k(M, \mathbf{R})$ contains a unique harmonic representative. For a differential form $\omega \in H^k(M, \mathrm{R})$, the Laplacian is defined by $\Delta\omega = \mathrm{dd}^*\omega + \mathrm{d}^*\mathrm{d}\omega$, where d^* is the codifferential operator characterized by $\int_M \langle \mathrm{d}^*\omega, \eta\rangle \,\mathrm{d}\mu(g) = \int_M \langle \omega, \mathrm{d}\eta\rangle \,\mathrm{d}\mu(g)$ for all $(k-1)$-forms η. A *harmonic form* ω is one which satisfies the elliptic equation $\Delta\omega = 0$, equivalently, using the compactness of M, harmonic forms are characterised by the equations $\mathrm{d}\omega \equiv \mathrm{d}^*\omega \equiv 0$. This remarquable theorem gives a direct link between the topology of M and solutions to an elliptic differential equation.

Harmonic mappings were introduced in 1954 by Sampson in an unpublished MIT report, in an attempt to obtain a homotopy version of Hodge theory. Another definition was proposed by J. Nash and H.E. Rauch (unpublished). F.B. Fuller (1954) gave the same definition as Sampson and obtained the first variation formula [7].

For a map $\varphi : (M^m, g) \to (N^n, h)$ between Riemannian manifolds, we define the *energy density* by $e(\varphi) = \frac{1}{2}|\mathrm{d}\varphi|^2$, where $|\mathrm{d}\varphi|^2 = \mathrm{Tr}_g \varphi^* h$ is the Hilbert-Schmidt norm of $\mathrm{d}\varphi$. If M is compact, we can now define the (total) *energy* to be the integral

$$E(\varphi) = \int_M e(\varphi)\mathrm{d}\mu(g)\,.$$

The Euler-Lagrange equations associated to this functional are

$$-\mathrm{d}^*\mathrm{d}\varphi := \tau(\varphi) = 0\,.$$

We note that, since automatically $\mathrm{dd}\varphi \equiv 0$, on a compact manifold, the Euler-Lagrange equations are equivalent to $\Delta \mathrm{d}\varphi \equiv 0$, where Δ is now the Laplacian acting on vector bundle valued forms (viewing $\mathrm{d}\varphi \in C^\infty(T^*M \otimes \varphi^{-1}TN)$). By analogy with Hodge theory, we can formulate the fundamental problem of harmonic maps as follows. *Given a homotopy class $\mathcal{H} \in C^\infty(M, N)$ of smooth maps from (M, g) to (N, h), find a harmonic representative $\varphi \in \mathcal{H}$* (We avoid discussion of the finer points of regularity theory in this introduction and suppose all maps smooth.)

The first major study of harmonic maps between Riemannian manifolds was made by Eells and Sampson in 1964 [5]. Their most striking theorem was the following: *Let (M^m, g) and (N^n, h) be compact Riemannian manifolds and suppose that (N, h) has non-positive sectional curvature. Then, for any homotopy class $\mathcal{H} \in C^\infty(M, N)$, there is a harmonic representative $\varphi \in \mathcal{H}$.*

As important as the result, was the method of proof. This was to use the heat equation:

$$\frac{\partial \varphi}{\partial t} = \tau(\varphi)\,. \tag{4}$$

The technique encapsulates the intuition referred to in the opening paragraph. We may imagine a closed curve sliding down the tube of a trombone until, at time infinity, it arrives at an absolute minimum of the energy where the tube is most narrow. The resulting map will be a geodesic – a special case of a harmonic map. In a similar way in which solitons occur (self-similar solutions), "bubbles" can occur in the harmonic map flow. This is studied in the article of Topping [23] and appears in a more general context in the contribution of Hebey [13]. The natural objects with which to describe this type of phenomena are called *rectifiable scans.* The setting is now the domain of geometric measure theory, and the application of these to Plateau-type minimisation problems is discussed in the contribution of De Pauw and Hardt [3].

It is known that some homotopy classes of maps cannot contain a harmonic representative, for example, there can be no harmonic map of degree 1 from the 2-torus to the 2-sphere [6]. However, a striking result of Eells and Ferreira, established in 1991 [4] realises Sampson's original goal, provided we allow deformations of the metric on the domain (so reinforcing our idea of interdependence between best metrics and best maps): *given two compact Riemannian manifolds* (M,g) *and* (N,h) *with* $\dim M \geq 3$ *and a homotopy class* $\mathcal{H}$ *of smooth maps between them, there is a smooth metric* $\tilde{g}$ *conformally equivalent to* g *and a map* $\varphi \in \mathcal{H}$ *such that* $\varphi : (M,\tilde{g}) \to (N,h)$ *is harmonic.* Their method of proof was to consider the modified energy functional

$$\frac{1}{2}\int_M (1+|\mathrm{d}\varphi|^2)^{p/2}\mathrm{d}\mu(g)\,. \tag{5}$$

Now, if $p > 2\dim M$, general principles from the direct method in the calculus of variations show that this attains its minimum in each homotopy class.

The appearance of the functional (5) illustrates the importance of the study of more general types of functionals. The most natural generalisation of the energy of a map, is the *p-energy*, defined by

$$E_p(\varphi) = \frac{1}{p}\int_M |\mathrm{d}\varphi|^p \mathrm{d}\mu(g)\,.$$

We may for example hope that special properties of harmonic maps in dimension 2 may be transferred to dimension p for critical points of this functional (*p-harmonic maps*). However, the Euler-Lagrange equations $\tau_p(\varphi) \equiv \mathrm{d}^*(|\mathrm{d}\varphi|^{p-2}\mathrm{d}\varphi) = 0$ degenerate when $\mathrm{d}\varphi = 0$ and the existence and regularity properties of p-harmonic maps becomes a difficult domain of study. The article of Hungerbühler [14] in this volume investigates the existence and regularity of p-harmonic maps by means of the heat flow.

Other aspects of harmonic maps, notably harmonic maps in *complex Finsler geometry* and harmonic maps from a *flat complex* are discussed in the articles of S. Nishikawa [18] and C. Mese [17], respectively, in this volume.

Finally, in the spirit of the evolution methods discussed above, one of the most successful programs in recent years has been the evolution of a hypersurface by its mean curvature, the so-called *mean curvature flow*. The article by C. Mantegazza [16] discusses the evolution of networks in the plane. This is an extension of these ideas in a new and imaginative way and represents the most recent work in this area.

The papers in this volume indicate the progress that has been made in recent years towards the goal of understanding the interplay between the topology and the geometry of manifolds. They are written by experts in their respective fields and we hope will provide inspiration and a platform for others to tackle some of the many beautiful problems that still remain.

References

[1] T. Aubin, *Equations différentielles non linéaires et problème de Yamabe concernant la courbure scalaire*, J. Math. Pures et Appl., **55** (1976), 269–296.

[2] P. Baird, A. Fardoun and R. Regbaoui, *The evolution of the scalar curvature of a surface to a prescribed function*, to appear, Ann. Scuola Norm. Sup. Pisa.

[3] T. De Pauw and R. Hardt, *Applications of scans and fractional power integrands*, this volume.

[4] J. Eells and M.J. Ferreira, *On representing homotopy calsses by harmonic maps*, Bull. London Math. Soc., **23**, (1991) 160–162.

[5] J. Eells and J.H. Sampson, *Harmonic mappings of Riemannian manifolds*, Amer. J. Math., **86**, (1964) 109–160.

[6] J. Eells and J.C. Wood, *Restrictions of harmonic maps of surfaces*, Topology, **15**, (1976) 263–266.

[7] F.B. Fuller, *Harmonic mappings*, Proc. Nat. Acad. Sci. U.S.A., **40**, (1954) 987–991.

[8] A. Gastel and M. Kronz, *A family of expanding Ricci solitons*, this volume.

[9] R. Hamilton, *Three-manifolds with positive Ricci curvature*, J. Differential Geom., **17** (1982), 255–306.

[10] R. Hamilton, *The Ricci flow on surfaces*, Contemp. Math., **71**, (1988) 237–262.

[11] R. Hardt and T. Rivière, *Connecting topological Hopf singularities*. To appear in Annali Sc. Norm. Sup. Pisa.

[12] E. Hebey, *Scalar curvature type problems in Riemannian geometry*, Notes of a course given at the University of Rome 3. Available at http://www.u-cergy.fr/rech/pages/hebey/.

[13] E. Hebey, *Bubbles over bubbles: A C^0-theory for the blow-up of second order elliptic equations of critical Sobolev growth*, this volume.

[14] N. Hungerbühler, *Heat flows into spheres for a class of energies*, this volume.

[15] D. Knopf, *Singularity models for the Ricci flow: an introductory survey*, this volume.

[16] C. Mantegazza, *Evolution by curvature of networks of curves in the plane*, this volume.

[17] C. Mese, *Regularity of harmonic maps from a flat complex*, this volume.

[18] S. Nishikawa, *Harmonic maps in complex Finsler geometry*, this volume.

[19] R. Schoen, *Conformal deformations of a Riemannian metric to constant scalar curvature*, J. Diff Geom., **20**, (1984) 479–495.

[20] P. Scott, *The geometries of 3-manifolds*, Bull. London Math. Soc., **15**, (1983) 401–487.

[21] C-L. Siegel, Topics in Complex Function Theory, Vol. 1, Wiley-Interscience, New York, 1969.

[22] H. Schwetlick and M. Struwe, *Convergence of the Yamabe flow for "large" energies*, to appear.

[23] P. Topping, *Bubbling of almost harmonic maps between 2-spheres at points of zero energy density*, this volume.

[24] N.S. Trudinger, *Remarks concerning the conformal deformation of Riemannian structures on compact manifolds*, Ann. Scuola Norm. Sup. Pisa (3), **22**, (1968) 265–274.

[25] R.O. Wells, Differential Analysis on Complex Manifolds, Springer Graduate Texts **65**, 1980.

[26] H. Yamabe, *On a deformation of Riemannian structures on compact manifolds*, Osaka Math. J., **12**, (1960) 21–37.

Paul Baird, Ali Fardoun and Rachid Regbaoui
Département de Mathématiques
Université de Bretagne Occidentale
UFR Sciences et Techniques
6 Av. Le Gorgeu
C.S, 93837
F-29238 Brest Cedex 3, France

Ahmad El Soufi
Laboratoire de Mathématiques et Physique Théorique
UMR 6083 du CNRS
Université de Tours
Parc de Grandmont
F-37200 Tours, France

Part I

Bubbling Phenomena

Progress in Nonlinear Differential Equations
and Their Applications, Vol. 59, 3–17

Bubbles over Bubbles: A C^0-theory for the Blow-up of Second Order Elliptic Equations of Critical Sobolev Growth

Emmanuel Hebey

Abstract. Let (M, g) be a smooth compact Riemannian manifold of dimension $n \geq 3$, and $\Delta_g = -\text{div}_g\nabla$ be the Laplace-Beltrami operator. Let also $2^\star$ be the critical Sobolev exponent for the embedding of the Sobolev space $H_1^2(M)$ into Lebesgue's spaces, and h be a smooth function on M. Elliptic equations of critical Sobolev growth like

$$\Delta_g u + hu = u^{2^\star - 1}$$

have been the target of investigation for decades. A very nice H_1^2-theory for the asymptotic behaviour of solutions of such an equation is available since the 1980's. We discuss here the C^0-theory recently developed by Druet, Hebey and Robert. Such a theory provides sharp pointwise estimates for the asymptotic behaviour of solutions of the above equation.

Let (M, g) be a smooth compact Riemannian manifold of dimension $n \geq 3$. We denote by $H_1^2(M)$ the standard Sobolev space of order two for integration and order one for differentiation. We let (h_α) be a sequence of $C^{0,\theta}$ functions on M, $0 < \theta < 1$, and consider equations like

$$\Delta_g u + h_\alpha u = u^{2^\star - 1} \tag{1}$$

where $\Delta_g = -\text{div}_g\nabla$ is the Laplace-Beltrami operator, $2^\star = 2n/(n-2)$ is the critical Sobolev exponent for the embedding of the Sobolev space $H_1^2(M)$ into Lebesgue's spaces, and u is required to be positive. We regard (1) as a possible model for more general second order elliptic equations of critical Sobolev growth. We let (u_α) be a bounded sequence in $H_1^2(M)$ of solutions of (1) in the sense that for any α,

$$\Delta_g u_\alpha + h_\alpha u_\alpha = u_\alpha^{2^\star - 1}$$

and $\|u_\alpha\|_{H_1^2} \leq \Lambda$ where $\Lambda > 0$ is independent of α. We also assume that the h_α's are uniformly bounded and that they converge in L^2 to some limiting function h_∞. Then, thanks to Struwe [32], we know how to describe the asymptotic behaviour of the u_α's in $H_1^2(M)$ as $\alpha \to +\infty$. More precisely, it follows from Struwe [32] that, up to a subsequence,

$$u_\alpha = u^0 + \sum B_\alpha^i + R_\alpha \tag{2}$$

where u^0 is a solution of the limit equation

$$\Delta_g u + h_\infty u = u^{2^\star-1}$$

the sum in the right-hand side of (2) is a finite sum over i, B^i_α is a bubble, obtained by rescaling fundamental positive solutions of the Euclidean equation $\Delta u = u^{2^\star-1}$, and the R_α's are lower order terms which converge strongly to 0 in $H^2_1(M)$. This asymptotic description provides a very satisfactory H^2_1-theory for the asymptotic behaviour of solutions of equations like (1). Let us assume now that the h_α's converge $C^{0,\theta}$ to h_∞ for some $0 < \theta < 1$. An important issue in the study of equations like (1) is to get a theory in which the above asymptotic description holds also in the C^0-space, where pointwise estimates are involved. Such a C^0-theory was recently developed in Druet, Hebey and Robert [12, 13]. We discuss this theory in these notes. Important applications of the theory are when dealing with sharp Sobolev inequalities. Monographs on sharp constant problems are Druet-Hebey [10], and Hebey [17]. Other possible applications are when dealing with the energy function, see for instance Hebey [18], or when dealing with compactness results as in Schoen [28, 29]. Very nice applications of the theory in such directions are in Druet [9].

1. Background material for the Euclidean space

We assume that $n \geq 3$. Thanks to the Sobolev embedding theorem [31], we refer also to Gagliardo [15] and Nirenberg [23], there exists $K > 0$ such that for any smooth function u with compact support in $\mathbb{R}^n$,

$$\|u\|_{2^\star} \leq K\|\nabla u\|_2 \tag{3}$$

where $2^\star = 2n/(n-2)$ is as above. Another classical result is that if K_n stands for the sharp constant in (3), then

$$K_n = \sqrt{\frac{4}{n(n-2)\omega_n^{2/n}}}$$

where ω_n is the volume of the unit n-sphere. Possible references on sharp Sobolev inequalities, including the Riemannian case, are Druet-Hebey [10] and Hebey [17]. The sharp Euclidean Sobolev inequality reads as

$$\|u\|_{2^\star} \leq K_n\|\nabla u\|_2 \ .$$

The extremal functions for this inequality are known. Let u be the function given by

$$u(x) = \left(\frac{1}{1+\frac{|x|^2}{n(n-2)}}\right)^{\frac{n-2}{2}} \tag{4}$$

Then, see for instance Bliss [2], u is an extremal function for the sharp Euclidean Sobolev inequality. Moreover, as it was shown by Caffarelli, Gidas and Spruck [6], see also Obata [24], u is the only positive solution of the critical Euclidean equation

$$\Delta u = u^{2^\star - 1} \tag{5}$$

which is such that $u(0) = \max_{\mathbb{R}^n} u = 1$. All the positive solutions of (5) are then given by

$$u_a^\lambda(x) = \lambda^{\frac{n-2}{2}} u\big(\lambda(x-a)\big) \tag{6}$$

where λ is any positive real number, a is any point in the Euclidean space, and u is the fundamental solution of (5) given by (4). In other words, positive solutions of (5) come from rescalings of the fundamental solution u given by (4). As a remark, the energy $E(u) = \|u\|_{2^\star}$ of u is given by $E(u) = K_n^{-(n-2)/2}$.

2. Background material for the unit sphere

The case of the unit sphere is closely related to the case of the Euclidean space. Let (S^n, h) be the unit n-sphere, $n \geq 3$. Let S_h be the scalar curvature of h. Then S_h is constant and $S_h = n(n-1)$. We consider the conformal scalar curvature equation

$$\Delta_h u + \frac{n(n-2)}{4} u = u^{2^\star - 1}$$

so that, with respect to (1),

$$h_\alpha = \frac{n-2}{4(n-1)} S_h$$

for all α. A classical result is that positive solutions of this equation are known. Let $x_0 \in S^n$ and r be the distance to x_0. Then, for any $\beta > 1$,

$$u_\beta = \left(\frac{n(n-2)}{4}(\beta^2 - 1)\right)^{\frac{n-2}{4}} \big(\beta - \cos r\big)^{1-\frac{n}{2}}$$

is a solution of the above equation. The energy $E(u) = \|u\|_{2^\star}$ of u_β is given by

$$E(u_\beta) = K_n^{-(n-2)/2}$$

where K_n is the sharp constant as in the preceding section. A possible reference in book form on such topics is Hebey [17]. It is easily seen that $u_\beta \to 0$ in $C^0_{loc}(S^n \backslash \{x_0\})$ as $\beta \to 1$. On the other hand, $u_\beta(x_0) \to +\infty$ as $\beta \to 1$. Thus the u_β's develop a singularity at x_0 as $\beta \to 1$. We say that the u_β's blow-up at x_0. Let x_β and μ_β be given by

$$u_\beta(x_\beta) = \max_{x \in S^n} u_\beta(x) = \mu_\beta^{1-\frac{n}{2}} .$$

Then $x_\beta = x_0$ and$\mu_\beta = \sqrt{\frac{4(\beta-1)}{n(n-2)(\beta+1)}}$. In particular, $\mu_\beta \to 0$ as $\beta \to 1$. Let B_β be the function given by

$$B_\beta(x) = \left(\frac{\mu_\beta}{\mu_\beta^2 + \frac{d_h(x_\beta,x)^2}{n(n-2)}}\right)^{\frac{n-2}{2}} .$$

We refer to B_β, see the following sections, as the standard bubble with respect to x_β and μ_β. It is easily seen in this particular situation that there exists $C > 1$ such that

$$\frac{1}{C} B_\beta(x) \leq u_\beta(x) \leq C B_\beta(x)$$

for all $x \in S^n$ and all $\beta > 1$. Similarly, for any $\varepsilon > 0$, there exists $\delta_\varepsilon > 0$ such that

$$\frac{1}{1+\varepsilon} B_\beta(x) \leq u_\beta(x) \leq (1+\varepsilon) B_\beta(x)$$

for all $x \in B_{x_0}(\delta_\varepsilon)$ when $\beta > 1$ is close to 1.Let R_β be the function given by

$$u_\beta = B_\beta + R_\beta .$$

Then, another easy claim in this particular situation is that

$$R_\beta \to 0 \text{ in } H_1^2(S^n)$$

as $\beta \to 1$. On the other hand, in general, $R_\beta \not\to 0$ in $C^0(S^n)$ as $\beta \to 1$, and the R_β's are not even bounded in $L^\infty(S^n)$. Let us assume for instance that $n \geq 7$, and let(y_β) be a sequence of points in S^n such that $d_h(x_0, y_\beta)^2 = (\beta - 1)^{1+\varepsilon}$ where $\varepsilon > 0$ is such that $\varepsilon < \frac{n-6}{4}$. Then, $R_\beta(y_\beta) \to +\infty$ as $\beta \to 1$. A more complete information is as follows. When $n \geq 7$, the R_β's are not bounded in $L^\infty(S^n)$. When $n = 6$, the R_β's are bounded in $L^\infty(S^n)$, but they do not converge to 0 in $C^0(S^n)$ as $\beta \to 1$. When $n = 3, 4, 5$, the R_β's converge to 0 in $C^0(S^n)$ as $\beta \to 1$. Other examples of blowing-up sequences are in Druet-Hebey [11] and Druet-Hebey-Robert [12].

3. Blow-up theory in the H_1^2-Sobolev space

When dealing with Sobolev spaces, we do not really need the notion of exact solutions. The more general notion of a Palais-Smale sequence works also. Let (M, g) be a smooth compact Riemannian manifold of dimension $n \geq 3$, and (h_α) be a sequence of continuous functions on M. We consider equation (1), and assume that there exists $C > 0$, and a continuous function h_∞ on M such that the following holds:

(A1) for any α, and any $x \in M$, $|h_\alpha(x)| \leq C$, and

(A2) $h_\alpha \to h_\infty$ in $L^2(M)$

as $\alpha \to +\infty$. Then, of course, it follows from (A1) and (A2) that the h_α's converge to h_∞ in all the Lebesgue spaces $L^p(M)$, $p > 1$. In other words, we assume that the h_α's are bounded in $L^\infty(M)$ and that they converge in Lebesgue spaces to some limiting function h_∞. Let u be a function in $H_1^2(M)$. We define the functional I_g^α by

$$I_g^\alpha(u) = \frac{1}{2}\int_M \left(|\nabla u|^2 + h_\alpha u^2\right) dv_g - \frac{1}{2^\star}\int_M |u|^{2^\star} dv_g$$

so that I_g^α is the formal primitive of equation (1). We let (u_α) be a sequence of functions in the Sobolev space $H_1^2(M)$. By definition, we say that (u_α) is a Palais-Smale sequence for equation (1) if the two following conditions hold:

(i) $I_g^\alpha(u_\alpha)$ is bounded with respect to α, and
(ii) $DI_g^\alpha(u_\alpha) \to 0$ strongly

as $\alpha \to +\infty$. This well-known notion of a Palais-Smale sequence extends the notion of a sequence of solutions of equation (1), since for an exact solution, the differential of I_g^α is zero. The goal in this section is to characterize the behaviour as $\alpha \to +\infty$ of such Palais-Smale sequences. In order to do that, we need the important notion of a bubble. We let η_δ be a cut-off function in $\mathbb{R}^n$ such that $\eta_\delta(x) = 1$ if $|x| \le \delta$ and $\eta_\delta(x) = 0$ if $|x| \ge 2\delta$, where $0 < \delta < i_g/2$ and i_g is the injectivity radius of (M, g). We let (x_α) be a sequence of points in M, and (R_α) be a sequence of positive real numbers such that $R_\alpha \to +\infty$ as $\alpha \to +\infty$. We let also u_a^λ be a positive solution of the critical Euclidean equation (5), so that u_a^λ is given by (6). Then we define a bubble as a sequence (B_α) of functions given by an equation like

$$B_\alpha(x) = R_\alpha^{\frac{n-2}{2}} \eta_{\delta,x_\alpha}(x) u_a^\lambda\left(R_\alpha \exp_{x_\alpha}^{-1}(x)\right)$$

where $\exp_{x_\alpha}$ is the exponential map at x_α, and $\eta_{\delta,x_\alpha}(x) = \eta_\delta\left(\exp_{x_\alpha}^{-1}(x)\right)$. In particular, since we do know what u_a^λ is, we do know the exact shape of bubbles. Returning to our question, we want to characterize the behaviour as $\alpha \to +\infty$ of a Palais-Smale sequence for equation (1). We let

$$\Delta_g u + h_\infty u = u^{2^\star - 1} \tag{7}$$

be the limit equation we get when we replace h_α by h_∞ in (1). The result which answers our question is as follows. Related references are Brézis-Coron [3, 4], Lions [22], Sacks-Uhlenbeck [27], Schoen [29], and Wente [33].

Theorem 3.1 (Struwe, [32]). *Assume* (A1) *and* (A2). *Let* (u_α) *be a Palais-Smale sequence for* (E_α), *with* $u_\alpha \ge 0$ *for all* α. *Then there exists some integer* k, *there exists a nonnegative solution* u^0 *of the limit equation* (7), *and there exist* k *bubbles* (B_α^i), i *running from* 1 *to* k, *such that, up to a subsequence,*

$$u_\alpha = u^0 + \sum_{i=1}^{k} B_\alpha^i + R_\alpha$$

for all α, *where* $R_\alpha \to 0$ *in* $H_1^2(M)$ *as* $\alpha \to +\infty$.

According to this theorem, a Palais-Smale sequence for (1) splits into a solution of the limit equation, which may be zero, a sum of bubbles that come from solutions of the critical Euclidean equation (5), and lower order terms in the

H_1^2-Sobolev space. As an important remark, the energy splits also. Another formulation of this is that bubbles do not interact at the H_1^2-level. Then, noting that the energy of a bubble is somehow a constant independent of the bubble, we do get a classification of the energy levels for which compactness does not hold. This gives in turn conditions under which u^0 is not the zero function, and thus, conditions under which the limit equation (7) possesses nontrivial solutions.The key point in the proof of this result is that given a Palais-Smale sequence, there exists a bubble such that the following holds. Namely that the Palais-Smale sequence minus the bubble is a Palais-Smale sequence whose energy is, up to lower order terms, the energy of the original Palais-Smale sequence minus some fixed-dimensional constant corresponding to the energy of a bubble. Then, to some extent, k in the theorem is the number of bubbles we have to substract to the original Palais-Smale sequence in order to get a Palais-Smale sequence of small energy. Noting that we do have compactness when the energy is sufficiently small, this proves the theorem.

Theorem 3.1 provides a very nice and complete H_1^2-theory for the blow-up of an equation like (1). The question now is to develop a C^0-theory involving pointwise estimates. As already mentioned in the introduction, such a C^0-theory was recently developed in Druet, Hebey and Robert [12, 13]. This is the subject of the following section. The naive idea here is that exact solutions of an elliptic PDE should satisfy C^0-estimates and not only H_1^2-estimates. Among others, a difficulty we have to face is that bubbles will interact at the C^0-level.

4. Blow-up theory in the C^0-space – the result

We assume in what follows that there exists $h_\infty \in C^{0,\theta}(M)$ such that the operator $\Delta_g + h_\infty$ is coercive, and such that

(A3) $\quad h_\alpha \to h_\infty$ in $C^{0,\theta}(M)$

as $\alpha \to +\infty$. By definition, $\Delta_g + h_\infty$ is coercive if its energy controls the H_1^2-norm. This is automatically the case if h_∞ is a positive function. We let (u_α) be a sequence of solutions of equation (1) in the sense that for any α, u_α is a solution of (1). We assume that the u_α's are bounded in the Sobolev space $H_1^2(M)$. Then, since $H_1^2(M)$ is a reflexive space, there exists some function u^0 in $H_1^2(M)$ such that, up to a subsequence,

$$u_\alpha \rightharpoonup u^0 \text{ weakly in } H_1^2(M) \tag{8}$$

as $\alpha \to +\infty$. Clearly, this is easy to check, u^0 is a solution of the limit equation

$$\Delta_g u + h_\infty u = u^{2^\star - 1} . \tag{9}$$

Another simple remark we get from standard elliptic theory is that either $u_\alpha \to u^0$ in $C^{2,\theta}(M)$, in which case we do have our asymptotic description, or we do have that

(A4) $\quad \max u_\alpha \to +\infty$

as $\alpha \to +\infty$. In this case, blow-up occurs, and this is the difficult situation in which we have to characterize the asymptotic behaviour of the u_α's. The theorem which answers our question, providing the C^0-theory we mentioned above, is as follows.

Theorem 4.1 (Druet-Hebey-Robert, [12, 13]**).** *Let (u_α) be a sequence of solutions of* (1), *$u_\alpha > 0$. We assume that the u_α's are bounded in $H_1^2(M)$, and that* (A3) *and* (A4) *hold. Then there exist $k \in \mathbb{N}^\star$, converging sequences $(x_{i,\alpha})$ in M, and sequences $(\mu_{i,\alpha})$ of positive real numbers converging to 0, $i = 1, \dots, k$, such that, up to a subsequence,*

$$(1-\varepsilon_\alpha)\,u^0(x) + \frac{1}{C}\sum_{i=1}^{k}\left(\frac{\mu_{i,\alpha}}{\mu_{i,\alpha}^2 + \frac{d_g(x_{i,\alpha},x)^2}{n(n-2)}}\right)^{\frac{n-2}{2}}$$

$$\le u_\alpha(x) \le (1+\varepsilon_\alpha)\,u^0(x) + C\sum_{i=1}^{k}\left(\frac{\mu_{i,\alpha}}{\mu_{i,\alpha}^2 + \frac{d_g(x_{i,\alpha},x)^2}{n(n-2)}}\right)^{\frac{n-2}{2}}$$

for all $x \in M$ and all α, where u^0 is the solution of the limit equation (9) *given by* (8), *$C > 1$ is independent of α and x, and (ε_α), independent of x, is a sequence of positive real numbers converging to 0 as $\alpha \to +\infty$. In particular, the u_α's are C^0-controlled, on both sides, by u^0 and standard bubbles.*

A complement to Theorem 4.1 is that C can be chosen as close as we want to 1 if we restrict the equation in Theorem 0.1 to small neighbourhoods of the geometrical blow-up points, defined as the limits of the $x_{i,\alpha}$'s. For instance, if $u^0 \not\equiv 0$, or if the u_α's just have one geometrical blow-up point, then for any $\varepsilon > 0$, there exists $\delta_\varepsilon > 0$ such that, up to a subsequence,

$$(1-\varepsilon_\alpha)\,u^0(x) + \frac{1}{1+\varepsilon}\sum_{i=1}^{k}\left(\frac{\mu_{i,\alpha}}{\mu_{i,\alpha}^2 + \frac{d_g(x_{i,\alpha},x)^2}{n(n-2)}}\right)^{\frac{n-2}{2}}$$

$$\le u_\alpha(x) \le (1+\varepsilon_\alpha)\,u^0(x) + (1+\varepsilon)\sum_{i=1}^{k}\left(\frac{\mu_{i,\alpha}}{\mu_{i,\alpha}^2 + \frac{d_g(x_{i,\alpha},x)^2}{n(n-2)}}\right)^{\frac{n-2}{2}}$$

for all α, all $x_0 \in \mathcal{S}$, and all $x \in B_{x_0}(\delta_\varepsilon)$, where $\mathcal{S}$ is the set consisting of the limits of the $x_{i,\alpha}$'s as $\alpha \to +\infty$, and $B_{x_0}(\delta_\varepsilon)$ is the geodesic ball of center x_0 and radius δ_ε. Outside the $B_x(\delta_\varepsilon)$'s, $x \in \mathcal{S}$, the u_α's converge $C^{2,\theta}$ to u^0. The estimate then extends to M in the particular case where $u^0 \not\equiv 0$. A refined estimate on the u_α's is given below in Section 5. Another complement to Theorem 4.1 is that the bubbles in this theorem satisfy the H_1^2-decomposition [32]. More precisely, we also have that for any $x \in M$ and any α,

$$u_\alpha(x) = u^0(x) + \sum_{i=1}^{k} \left(\frac{\mu_{i,\alpha}}{\mu_{i,\alpha}^2 + \frac{d_g(x_{i,\alpha},x)^2}{n(n-2)}} \right)^{\frac{n-2}{2}} + R_\alpha(x)$$

where the R_α's, $R_\alpha \in H_1^2(M)$ for all α, are such that $R_\alpha \to 0$ strongly in $H_1^2(M)$ as $\alpha \to +\infty$. Moreover, if we define the energy of a function u by $E(u) = \|u\|_{2^\star}$, then

$$E(u_\alpha)^{2^\star} = E(u^0)^{2^\star} + kK_n^{-n} + r_\alpha$$

for all α, where the r_α's, $r_\alpha \in \mathbb{R}$, are such that $r_\alpha \to 0$ as $\alpha \to +\infty$. In other words, the energies split also.

Atkinson-Peletier [1] and Brezis-Peletier [5] have been concerned with the description of the pointwise behaviour of sequences of solutions of equations like (1), dealing with radially symmetrical solutions u_ε of the semi-critical equations $\Delta u = u^{2^\star-1-\varepsilon}$ on the unit ball of the Euclidean space. More recent developments in this specific direction are in Robert [25, 26]. An estimate like in Theorem 4.1 in the case $k = 1$, stating that solutions of minimal energy of equations like (1) are controlled from above by a standard bubble, appeared then in Han [16] when dealing with solutions u_ε of the equations $\Delta u = u^{2^\star-1-\varepsilon}$ on bounded open subsets of the Euclidean space, in Hebey-Vaugon [19] when dealing with (1) and arbitrary Riemannian manifolds, and in Li [20, 21] when dealing with equations like (1) on the unit sphere. Improvements, still in the case $k = 1$, are in Druet [7, 8] and Druet-Robert [14]. We refer also to Schoen-Zhang [30].

5. Proof of the theorem in the general case

We only present a very brief sketch of the proof of Theorem 4.1 when we regard the theorem under its entire generality. We assume in what follows that

$$\max u_\alpha \to +\infty$$

as $\alpha \to +\infty$. If not the case, up to a subsequence, the u_α's converge $C^{2,\theta}$ to u^0, and we do have our asymptotic description. As a remark, the proof of the theorem goes through the proof of a slightly stronger result. All what follows is up to a subsequence. We let G be the Green function of the operator $\Delta_g + h_\infty$, and let Φ be the continuous function on $M \times M$ given by

$$\Phi(x,y) = (n-2)\omega_{n-1} d_g(x,y)^{n-2} G(x,y)$$

if $x \neq y$, and

$$\Phi(x,y) = 1$$

if $x = y$, where ω_{n-1} is the volume of the unit $(n-1)$-sphere. Then we claim that for any converging sequence (x_α) of points in M, and any α,

$$u_\alpha(x_\alpha) = \Big(1 + o(1)\Big)u^0(x_\alpha) + \sum_{i=1}^{k}\Big(\Phi(x_i, x) + o(1)\Big)B_\alpha^i(x_\alpha)$$

where x is the limit of the x_α's, x_i is the limits of the $x_{i,\alpha}$'s, and B_α^i is the standard bubble with respect to the $x_{i,\alpha}$'s and $\mu_{i,\alpha}$'s given by Theorem 4.1. It is easily seen that the theorem and the remarks after this theorem follow from such an asymptotic description. The proof of these asymptotics splits into four main steps. The first preliminary step consists in getting rescaling invariant estimates. Such estimates basically state that there exist $k \in \mathbb{N}^\star$, converging sequences $(x_{i,\alpha})$ in M, and sequences $(\mu_{i,\alpha})$ of positive real numbers converging to 0, $i = 1, \dots, k$, such that

$$R_\alpha^k(x)^{\frac{n}{2}-1}u_\alpha(x) \le C$$

for all x and all α, where$R_\alpha^k(x)$ is the minimum over i of the distances from the $x_{i,\alpha}$'s to x. The key idea here is that if such an estimate is false, then we can construct another blow-up point. This weak estimate comes with an important refinement and complementary informations on the limit of the u_α's outside blow-up points. Then we need to prove that the upper estimate in Theorem 4.1 holds. For that purpose, we rearrange the $x_{i,\alpha}$'s in families. Inside a family, blow-up points are close one to the other. Two families are far one from the other. If the $y_{i,\alpha}$'s are the representative of such families having the largest $\mu_{i,\alpha}$, $i = 1, \dots, p$, the second step in the proof consists in proving that an upper estimate like in Theorem 4.1 holds with respect to the $y_{i,\alpha}$'s. We prove that there exist $C > 0$ and $R > 0$ such that

$$u_\alpha(x) \le \big(1 + o(1)\big)u^0(x) + C\sum_{i=1}^{p} B_\alpha^i(x)$$

for all α and all $x \in M\backslash \cup_{i=1}^{p} B_{y_{i,\alpha}}(R\mu_{i,\alpha})$, where B_α^i is the standard bubble with respect to $y_{i,\alpha}$ and its corresponding $\mu_{i,\alpha}$. The proof of such an estimate goes through the establishment of a scale of intermediate estimates, referred to as ε-sharp estimates with $0 < \varepsilon < \frac{n-2}{2}$, the weakest of these when $\varepsilon = \frac{n-2}{2}$ being like the weak estimate we discussed above. The proof of the upper estimate as in the theorem then reduces to the proof that this estimate holds inside the $B_{y_{i,\alpha}}(R\mu_{i,\alpha})$'s. This is the third step in the proof. We proceed here by induction. We consider subfamilies of blow-up points, and prove that the estimate holds in $B_{y_{i,\alpha}}(R\mu_{i,\alpha})$, outside smaller balls of sub-representatives, and so on up to the point where we have exhausted all the blow-up points. In each step of this induction process, we pass through ε-sharp estimates, and, in some sense, let then $\varepsilon \to 0$ to get the sharp estimate. Once we have proved that the upper estimate of Theorem 4.1 holds in M, the argument becomes simplier and more conventional. The fourth step in the proof consists in proving that the above asymptotics follow from the

Green representation formula

$$\begin{aligned} u_\alpha(x_\alpha) - u^0(x_\alpha) &= \int_M G_\alpha(x_\alpha, x)\big(u_\alpha(x)^{2^\star-1} - u^0(x)^{2^\star-1}\big)dv_g \\ &\quad + \int_M G_\alpha(x_\alpha, x)\big(h_\infty(x) - h_\alpha(x)\big)u^0(x)dv_g \end{aligned}$$

where G_α is the Green function for the operator $\Delta_g + h_\alpha$. The different terms that are involved in this formula are controlled thanks to the upper estimate we have just discussed. The asymptotics follow from rather standard developments. As already mentioned, since Φ is continuous, the theorem and the remarks after the theorem are then easy consequences of the asymptotics. We refer to Druet, Hebey and Robert [12] for a complete proof, and also to the next section for the special case where $k = 1$.

6. Proof of the theorem when the energy is of minimal type

As a preliminary remark, it is easily checked that the energy of a bubble is a constant which is independent of the bubble. More precisely, if B_α is a bubble, then

$$E(B_\alpha) = K_n^{-\frac{n-2}{2}} + o(1)$$

where $o(1) \to 0$ as $\alpha \to +\infty$, and K_n is the sharp constant for the Sobolev inequality on the Euclidean space. Returning to equation (E_α), we let (u_α) be a sequence of solutions of (E_α). We assume that for any α,

$$E(u_\alpha) \le K_n^{-(n-2)/2} \,. \tag{10}$$

The theory in this particular situation is due to Druet and Robert. A complete proof can be found in Druet and Hebey [10]. Thanks to (10), the u_α's are bounded in $H_1^2(M)$. It follows that there exists $u^0 \in H_1^2(M)$ such that, up to a subsequence,$u_\alpha \rightharpoonup u^0$ in $H_1^2(M)$ as $\alpha \to +\infty$. Since the energy of the u_α's is less than or equal to the minimum energy, see (10), the following holds:

(P1) either $u^0 \not\equiv 0$, and $u_\alpha \to u^0$ in $C^{2,\theta}(M)$ as $\alpha \to +\infty$,

(P2) or $u^0 \equiv 0$, and the u_α's blow-up with one bubble.

The general idea here is that we do not have enough energy so that we can face a multiple bubbles situation, neither the situation where we would have a nonzero limit u^0 and a bubble. Now the statement we want to prove is the following. We assume that $u^0 \equiv 0$, if not we do have our asymptotic description, and we assume that (10) holds. Then we want to prove that there exists some constant $C > 1$, that there exists a converging sequence (x_α) of points in M, and that there exists a sequence (μ_α) of real numbers, $\mu_\alpha > 0$ and $\mu_\alpha \to 0$ as $\alpha \to +\infty$, such that, up to a subsequence,

$$\frac{1}{C}\left(\frac{\mu_\alpha}{\mu_\alpha^2+\frac{d_g(x_\alpha,x)^2}{n(n-2)}}\right)^{\frac{n-2}{2}} \le u_\alpha(x) \le C\left(\frac{\mu_\alpha}{\mu_\alpha^2+\frac{d_g(x_\alpha,x)^2}{n(n-2)}}\right)^{\frac{n-2}{2}} \tag{11}$$

for all x, and all α. This is exactly Theorem 4.1 in this particular situation where the energy is assumed to be of minimal type. In such a particular situation, we can define explicitely the x_α's and the μ_α's. More precisely, the x_α's and μ_α's are here given by

$$u_\alpha(x_\alpha) = \max_M u_\alpha = \mu_\alpha^{1-\frac{n}{2}} . \tag{12}$$

In other words, x_α is a point where u_α is maximum, and μ_α is, up to some power, the inverse of the maximum of u_α. As when discussing the general case, we need first to prove the upper estimate in our statement, namely in (11). A simple claim then, thanks to (12), is that this upper estimate is equivalent to the following estimate: there exists $C > 1$ such that, up to a subsequence,

$$d_g(x_\alpha,x)^{n-2}\mu_\alpha^{-\frac{n-2}{2}}u_\alpha(x) \le C \tag{13}$$

for all x, and all α. In order to prove (13) we proceed in three steps. All what follows is up to a subsequence. As above, the first step consists in proving rescaling invariant estimates. To some extent, these estimates state that the u_α's are controlled by the square root of a bubble. They can be stated as follows:

Step 1 (Rescaling invariant estimates): There exists $C > 0$ such that for any α and any x,

$$d_g(x_\alpha,x)^{\frac{n-2}{2}}u_\alpha(x) \le C . \tag{14}$$

Moreover, for any $R > 0$, for any α, and for any $x \in M\backslash B_{x_\alpha}(R\mu_\alpha)$,

$$d_g(x_\alpha,x)^{\frac{n-2}{2}}u_\alpha(x) \le \varepsilon_{R,\alpha} \tag{15}$$

where $\varepsilon_{R,\alpha}$ is such that $\lim\limits_{R\to+\infty}\lim\limits_{\alpha\to+\infty}\varepsilon_{R,\alpha} = 0$.

We have here two estimates. The second estimate is a refinement of the first one. The general idea of the proof of (14) and (15) is that if one of these two estimates is false, then we must have another bubble in the decomposition of the u_α's. This is not possible since we assumed that the energy of the u_α's is of minimal type, so that we just have one single bubble in the decomposition. In Step 2, we prove that the upper estimate in our statement (11) holds up to some ε, $\varepsilon > 0$. In other words, we prove something like an ε-sharp upper estimate in (11).

Step 2 (ε-sharp upper estimate): For any $\varepsilon > 0$, there exists $C_\varepsilon > 0$,such that

$$d_g(x_\alpha,x)^{n-2-\varepsilon}\mu_\alpha^{-\frac{n-2}{2}+\varepsilon}u_\alpha(x) \le C_\varepsilon \tag{16}$$

for all x, and all α.

As a remark, (14) gives that (16) holds with $\varepsilon = (n-2)/2$. Another remark is that (16) becomes stronger and stronger as ε decreases, $\varepsilon \to 0$. The proof of this estimate makes use of the existence of a Green function for the second order operator in the left-hand side of equation (1). What we actually prove, thanks to maximum principles, is that the u_α's are bounded from above by a $(1-\varepsilon)$-power of this Green function. This implies (16). In Step 3, we prove that the upper estimate in our statement is also true. In other words, we prove that we can take $\varepsilon = 0$ in the ε-sharp upper estimate (16).

Step 3 (Upper estimate): There exists $C > 0$ such that

$$d_g(x_\alpha, x)^{n-2}\mu_\alpha^{-\frac{n-2}{2}} u_\alpha(x) \le C \tag{17}$$

for all x, and all α.

As already mentioned, (17) is exactly the upper estimate of our statement (11), namely that the u_α's are bounded from above by some constant C times the standard bubble defined by the x_α's and μ_α's. The proof of this estimate goes through the Green representation formula, the u_α's being controlled thanks to the ε-sharp upper estimate (16), $\varepsilon > 0$ small. With such an estimate we have half of our statement. Summarizing Steps 1–3, we prove first that we do have rescaling invariant estimates. This corresponds to the case where $\varepsilon = (n-2)/2$ in (16). Then we prove that the ε-sharp upper estimates (16) hold when $0 < \varepsilon < (n-2)/2$. Then, at last, we prove that our upper estimate (17) holds. This corresponds to the case where $\varepsilon = 0$ in (16). Step 2 uses Step 1. Step 3 uses Step 2. As a remark, we assumed here that the h_α's converge in $C^{0,\theta}$ to some limit function h_∞. The rescaling invariant estimates actually hold if we only assume that the negative part of h_α is small in some L^p-norm. The ε-sharp upper estimate and the upper estimate hold if we only assume that the h_α's are greater than or equal to some function V for which the operator $\Delta_g + V$ is coercive.

Now that we have the upper estimate in (11), we can discuss the last step in the proof. We let B_α be the standard bubble in (11), and G be the Green function of the operator $\Delta_g + h_\infty$. We let also x_0 be the geometrical blow-up point, defined as the limit of the x_α's as $\alpha \to +\infty$. At last, as in the preceding section, we let $\Phi : M \times M \to \mathbb{R}$ be given by $\Phi(x,y) = (n-2)\omega_{n-1} d_g(x,y)^{n-2} G(x,y)$ if $x \neq y$, and $\Phi(x,y) = 1$ if $x = y$, where ω_{n-1} is the volume of the unit $(n-1)$-sphere. A simple claim, which follows from the construction of the Green function, is that Φ is continuous on $M \times M$. Step 4 in the proof can then be stated as follows:

Step 4 (Asymptotic behaviour): for any converging sequence $(\tilde{x}_\alpha)$ of points in M,

$$\lim_{\alpha \to +\infty} \frac{u_\alpha(\tilde{x}_\alpha)}{B_\alpha(\tilde{x}_\alpha)} = \Phi(x_0, \tilde{x}_0)$$

where $\tilde{x}_0$ is the limit of the $\tilde{x}_\alpha$'s as $\alpha \to +\infty$.

We prove Step 4 by using once again the Green representation formula, the u_α's being controlled thanks to the upper estimate of Step 3. As already mentioned, our conclusion easily follows from Step 4. Noting that Φ is positive, and since Φ is continuous, we get with Step 4 that there exists $C > 1$ such that

$$\frac{1}{C} B_\alpha(x) \le u_\alpha(x) \le C B_\alpha(x)$$

for all x, and all α. This proves (11), and thus Theorem 4.1 in this particular situation where the energy is assumed to be of minimal type. Moreover, since $\Phi = 1$ on the diagonal, and since Φ is continuous, we also get with Step 4 that for any $\varepsilon > 0$, there exists $\delta_\epsilon > 0$ such that

$$\frac{1}{1+\varepsilon} B_\alpha(x) \le u_\alpha(x) \le (1+\varepsilon) B_\alpha(x)$$

for all $x \in B_{x_0}(\delta_\varepsilon)$, and for all α. This proves, in this particular situation where the energy is assumed to be of minimal type, the remark after Theorem 4.1 concerning the sharpness of the constant.

References

[1] Atkinson, F.V., and Peletier, L.A., *Elliptic equations with nearly critical growth*, J. Diff. Eq., 70, 349–365, 1987.

[2] Bliss, G. A., *An integral inequality*, J. London Math. Soc., 5, 40–46, 1930.

[3] Brézis, H., and Coron, J.M., *Convergence de solutions de H-systèmes et applications aux surfaces à courbure moyenne constante*, C.R. Acad. Sci. Paris, 298, 389–392, 1984.

[4] Brézis, H., and Coron, J.M., *Convergence of solutions of H-systems or how to blow bubbles*, Arch. Rational Mech. Anal., 89, 21–56, 1985.

[5] Brézis, H., and Peletier, L.A., *Asymptotics for elliptic equations involving critical Sobolev exponents*, in Partial Differential Equations and the Calculus of Variations, eds. F. Colombini, A. Marino, L. Modica and S. Spagnalo, Basel: Birkhäuser, 1989.

[6] Caffarelli, L. A., Gidas, B., and Spruck, J., *Asymptotic symmetry and local behavior of semilinear elliptic equations with critical Sobolev growth*, Comm. Pure Appl. Math., 42, 271–297, 1989.

[7] Druet, O., *The best constants problem in Sobolev inequalities*, Math. Ann., 314, 327–346, 1999.

[8] ———, *Sharp local isoperimetric inequalities involving the scalar curvature*, Proc. Amer. Math. Soc., 130, 2351–2361, 2002.

[9] ———, *From one bubble to several bubbles. The low-dimensional case*, J. Differential Geom., To appear.

[10] Druet, O., and Hebey, E., *The AB program in geometric analysis. Sharp Sobolev inequalities and related problems*, Memoirs of the American Mathematical Society, MEMO/160/761, 2002.

[11] Druet, O., and Hebey, E., *Blow-up examples for second order elliptic PDEs of critical Sobolev growth*, Trans. Amer. Math. Soc., to appear.

[12] Druet, O., Hebey, E., and Robert, F., *Blow-up theory for elliptic PDEs in Riemannian geometry*, Mathematical Notes, Princeton University Press, to appear.

[13] Druet, O., Hebey, E., and Robert, F., *A C^0-theory for the blow-up of second order elliptic equations of critical Sobolev growth*, Electronic Research Announcements of the AMS, 9, 19–25, 2003.

[14] Druet, 0., and Robert, F., *Asymptotic profile for the sub-extremals of the sharp Sobolev inequality on the sphere*, Comm. P.D.E., 25, 743–778, 2001.

[15] Gagliardo, E.,*Proprietà di alcune classi di funzioni in più variabili*, Ricerche Mat., 7, 102–137, 1958.

[16] Han, Z.C., *Asymptotic approach to singular solutions for nonlinearelliptic equations involving critical Sobolev exponent*, Ann. Inst. H. Poincaré. Anal. Non Linéaire, 8, 159–174, 1991.

[17] Hebey, E., *Nonlinear analysis on manifolds: Sobolev spaces and inequalities*, CIMS Lecture Notes, Courant Institute of Mathematical Sciences, Vol. 5, 1999. Second edition published by the American Mathematical Society, 2000.

[18] ———, *Nonlinear elliptic equations of critical Sobolev growth from a dynamical viewpoint*, Preprint, Conference in honor of H. Brézis and F. Browder, Rutgers University, 2001.

[19] Hebey, E., and Vaugon, M., *The best constant problem in the Sobolev embedding theorem for complete Riemannian manifolds*, Duke Math. J., 79, 235–279, 1995.

[20] Li, Y.Y., *Prescribing scalar curvature on S^n and related problems, Part I*, J. Diff. Equations, 120, 319–410, 1995.

[21] ———, *Prescribing scalar curvature on S^n and related problems, Part II: existence and compactness*, Comm. Pure Appl. Math., 49, 541–597, 1996.

[22] Lions, P.L., *The concentration-compactness principle in the calculus of variations. The limit case. I, II*,Rev. Mat. Iberoamericana, 1: no. 1, 145–201; no. 2, 45–121, 1985.

[23] Nirenberg, L., *On elliptic partial differential equations*, Ann. Scuola Norm. Sup. Pisa, 13, 116–162, 1959.

[24] Obata, M., *The conjectures on conformal transformations of Riemannian manifolds*, J. Differential Geom., 6, 247–258, 1971/72.

[25] Robert, F., *Asymptotic behaviour of a nonlinear elliptic equation with critical Sobolev exponent - The radial case I*, Adv. Differential Equations, 6, 821-846, 2001.

[26] ———, *Asymptotic behaviour of a nonlinear elliptic equation with critical Sobolev exponent - The radial case II*, Nonlinear Differ. Eq. Appl., 9, 361–384, 2002.

[27] Sacks, P., and Uhlenbeck, K., *On the existence of minimal immersions of 2-spheres*, Ann. of Math., 113, 1–24, 1981.

[28] Schoen, R., *Variational theory for the total scalar curvature functional for Riemannian metrics and related topics*, Topics in calculus of variations (Montecatini Terme, 1987), 120–154. Lecture Notes in Mathematics, 1365, Springer, Berlin-New York, 1989.

[29] ———, *On the number of constant scalar curvature metrics in a conformal class*, Differential Geometry: A symposium in honor of Manfredo Do Carmo, H.B.Lawson and K.Tenenblat eds., Pitman Monogr. Surveys Pure Appl. Math., 52, 311–320, 1991.

[30] Schoen, R., and Zhang, D., *Prescribed scalar curvature on the n-sphere*, Calc. Var. Partial Differential Equations, 4, 1–25, 1996.

[31] Sobolev, S.L., *Sur un théorème d'analyse fonctionnelle*, Mat. Sb., 46, 471–496, 1938.

[32] Struwe, M., *A global compactness result for elliptic boundary problems involving limiting nonlinearities*, Math. Z., 187, 511–517, 1984.

[33] Wente, H.C., *Large solutions to the volume constrained Plateau problem*, Arch. Rational Mech. Anal., 75, 59–77, 1980.

Emmanuel Hebey
Université de Cergy-Pontoise
Département de Mathématiques
Site de Saint-Martin
2 avenue Adolphe Chauvin
F-95302 Cergy-Pontoise cedex, France
e-mail: Emmanuel.Hebey@math.u-cergy.fr

Progress in Nonlinear Differential Equations
and Their Applications, Vol. 59, 19–31

Application of Scans and Fractional Power Integrands

Thierry De Pauw and Robert Hardt

Abstract. In this note we describe the notion of a rectifiable scan and consider some applications [DH1], [DH2] to Plateau-type minimization problems. "Scans" were first introduced in the work [HR1] of Tristan Rivière and the second author to adequately describe certain bubbling phenomena. There, the behaviour of certain $W^{1,3}$ weakly convergent sequences of smooth maps from four-dimensional domains into $\mathbf{S}^2$ led to the consideration of a necessarily infinite mass generalization of a rectifiable current. The definition of a scan is motivated by the fact that a rectifiable current can be expressed in terms of its lower-dimensional slices by oriented affine subspaces. By integral geometry, the slicing function for the rectifiable current is a mass integrable function of the subspaces. With a scan one considers more general such functions that are not necessarily mass integrable.

1. Rectifiable currents and the plateau problem

An m-dimensional *rectifiable set* R in $\mathbf{R}^n$ is a subset of some countable union $\cup_{i=0}^{\infty} M_i$ where $M_1, M_2, \ldots$ are m-dimensional $\mathcal{C}^1$ submanifolds and M_0 has m-dimensional Hausdorff measure $\mathcal{H}^m(M_0) = 0$. At $\mathcal{H}^m$ almost every point $x \in R$, R has an approximate tangent space. [S], Section 3. An m-dimensional (integer-multiplicity) *rectifiable current* T in $\mathbf{R}^n$ is given by a bounded m-dimensional Borel measurable rectifiable *concentration* set R_T together with an $\mathcal{H}^m$-integrable density function $\theta_T : R_T \to \{1, 2, \ldots\}$ and an $\mathcal{H}^m$-measurable orientation $\vec{T} : R_T \to \bigwedge_m \mathbf{R}^n$ so that at $\mathcal{H}^m$ almost every $x \in R_T$, $\vec{T}(x)$ is the wedge product of vectors from an orthonormal basis of the approximate tangent space of R_T at x. See [F1],4.1.24 or [S], Section 27. Thus the action of the current T on a differential m-form $\phi \in \mathcal{D}^m(\mathbf{R}^n)$ is given by the integration

$$T(\phi) = \int_{R_T} \langle \vec{T}(x), \phi(x) \rangle \, \theta_T(x) \, d\mathcal{H}^m x .$$

The *mass* of T is then simply $\mathbf{M}(T) = \int_{R_T} \theta_T(x) \, d\mathcal{H}^m x$. For $m \geq 1$, *boundary* of T is the $m-1$-dimensional current defined by the formula $\partial T(\psi) = T(d\psi)$ for $\psi \in \mathcal{D}^{m-1}(\mathbf{R}^n)$. A rectifiable current generalizes an oriented submanifold M. We sometimes use the abbreviated notation $[\![M]\!]$, in case the orientation is known, for the corresponding multiplicity one rectifiable current. So if M is an oriented manifold with boundary, Stokes Theorem becomes $\partial[\![M]\!] = [\![\partial M]\!]$. Even though $\mathcal{H}^m(R_T) < \infty$, one should be aware that the *support* of the current, $\operatorname{spt} T$, which may be much larger than R_T, might even be n-dimensional for $m \geq 1$. On the other hand a *zero*-dimensional rectifiable current is simply a finite integral combination of point masses. Let $\mathcal{R}_m$ denote the group of m-dimensional rectifiable currents in $\mathbf{R}^n$. In 1960, H. Federer and W. Fleming obtained the following fundamental existence theorem:

1.1 Theorem. [FF] *Given any $T_0 \in \mathcal{R}_m$ with $\partial T_0 \in \mathcal{R}_{m-1}$, the family of currents $\{T \in \mathcal{R}_m \ : \ \partial T = \partial T_0\}$ contains a rectifiable current of least mass.*

This theorem is valid for all $m \geq 1$ in any $\mathbf{R}^n$ as well as in any compact Riemannian manifold (provided the admissible family is nonempty). There one also has, in any homology class, a rectifiable current that minimizes mass. Among *general* currents the existence of a mass minimizers is an easy consequence of the Banach-Alaoglu theorem, but what is important in [FF] is the rectifiability, which should be understood as an initial regularity for minimizers. The complete interior regularity, of such rectifiable mass-minimizers, i.e., that $\operatorname{spt} T \setminus \operatorname{spt} \partial T$ is an embedded real analytic submanifold, was established in the sixties for $1 \leq m = n-1 \leq 6$ by works of Fleming [Fl], De Giorgi [D], Almgren [A1], Triscari [Tr], and Simons [Ss]. De Giorgi's work showed that a seven-dimensional mass-minimizer in $\mathbf{R}^8$ would have at most isolated interior singularities, and, in fact in 1970, Bombieri, De Giorgi [D], and Giusti [BDG] established the mass-minimality of the specific example

$$Q \ = \ \partial[\![\{(x,y) \in \mathbf{R}^4 \times \mathbf{R}^4 \ : \ |x| < |y|\}]\!] \llcorner \mathbf{B}_1^8 \ ,$$

which has an isolated singularity at $(0,0)$. Then the cartesian product with a cube $Q \times [\![-1,1]\!]^j$ is mass minimizing in $\mathbf{R}^{8+j}$, and so the following result of H. Federer gives the optimal estimate of the Hausdorff dimension of the interior singular set of a codimension one minimizer:

1.2 Theorem. [F2] *For any m-dimensional mass-minimizing rectifiable current T in $\mathbf{R}^{m+1}$ and $\epsilon > 0$*

$$\mathcal{H}^{m-7+\epsilon}\big(\operatorname{Sing}(\operatorname{spt} T \setminus \operatorname{spt} \partial T)\big) \ = \ 0 \ .$$

The complete boundary regularity of T was established by Hardt and Simon [HS] near any point where the given ∂T_0 is a smooth $m-1$-dimensional oriented embedded submanifold of $\mathbf{R}^{m+1}$. In 1984, F.J. Almgren completed a massive work treating the higher codimension interior partial regularity:

1.3 Theorem. [A4] *For any m-dimensional mass-minimizing rectifiable current T in $\mathbf{R}^n$ and $\epsilon > 0$*

$$\mathcal{H}^{m-2+\epsilon}\big(\operatorname{Sing}(\operatorname{spt} T \setminus \operatorname{spt} \partial T)\big) \ = \ 0 \ ,$$

and S. Chang [C] showed that interior singularities are at most isolated points for $m = 2$. Note that [FF] already contained the singular minimizing example of the sum of two oriented totally orthogonal disks in $\mathbf{R}^2 \times \mathbf{R}^2$.

$$[\![\mathbf{B}_1^2 \times \{0\}]\!] + [\![\{0\} \times \mathbf{B}_1^2]\!] .$$

2. Size and fractional powers of the density

In $\mathbf{R}^3$, two-dimensional mass-minimizing rectifiable currents have no interior singularities and provide a nice model for some but not all "soap films". General soap films may have interior singular curves which simultaneously border three surfaces meeting at equal angles. To use currents in a better model for soap films, Almgren [A3] introduced the notion of *size* for a rectifiable current:

$$\text{Size}\,(T) = \mathcal{H}^m(R_T) .$$

Thus one ignores the density function in computing size. To understand size versus mass minimization, consider the one-dimensional rectifiable current in the plane consisting of 2 parallel similarly oriented intervals

$$T_0 = [\![(-\frac{1}{2}, \frac{\sqrt{3}}{2}), (\frac{7}{2}, \frac{\sqrt{3}}{2})]\!] + [\![(-\frac{1}{2}, -\frac{\sqrt{3}}{2}), (\frac{7}{2}, -\frac{\sqrt{3}}{2})]\!] .$$

Then T_0 is mass-minimizing among all rectifiable currents having boundary equal to ∂T_0. However, the size minimizer is sum of five oriented intervals,

$$\begin{aligned} T_1 &= [\![(-\frac{1}{2}, \frac{\sqrt{3}}{2}), (0,0)]\!] + [\![(-\frac{1}{2}, -\frac{\sqrt{3}}{2}), (0,0)]\!] \\ &+ 2[\![(0,0), (3,0)]\!] + [\![(3,0), (\frac{7}{2}, \frac{\sqrt{3}}{2})]\!] + [\![(3,0), (\frac{7}{2}, \frac{\sqrt{3}}{2})]\!] , \end{aligned}$$

one of which has multiplicity 2. Note that

$\mathbf{M}(T_0) = 2{\cdot}4 < 4{\cdot}1{+}2{\cdot}3 = \mathbf{M}(T_1)$ and $\text{Size}\,(T_0) = 8 > 4{\cdot}1{+}3 = \text{Size}\,(T_1)$.

Similarly in dimension 2 one may consider the sum T_0 of two close coaxial, parallel and similarly oriented, disks. Then T_0 is mass-minimizing, but the size-minimizer T_1 with boundary ∂T_0 contains a single multiplicity 2 disk in the middle and the set $\operatorname{spt} T_1$ models a soap film with an interior singular curve. A fundamental problem with size-minimization is the lack of a general existence theorem. One is faced with the possibility of a size-minimizing sequence of rectifiable currents having unbounded masses and failing to have subsequences convergent as currents. This is what happens in an example of F. Morgan.

2.1 Example [M]. For a fixed $1 < \beta < 2$ consider the following countable sum of vertical oriented intervals in the plane:

$$T_0 = \sum_{j=1}^{\infty} [\![(\frac{1}{j}, -\frac{1}{j^\beta}), (\frac{1}{j}, \frac{1}{j^\beta})]\!] .$$

Then $T_0 \in \mathcal{R}_1$ because $\mathbf{M}(T_0) = \sum_{j=1}^\infty \frac{2}{j^\beta} < \infty$. One may check that T_0 is mass-minimizing by an easy callibration argument. However, one sees, as in the previous section, that one may decrease the size, at the expense of increasing the mass, by replacing the oriented interval $[\![(1,-1),(1,1)]\!]$ by the sum of three intervals

$$[\![(1,-1),(\frac{1}{2},1)]\!] + [\![(\frac{1}{2},-\frac{1}{2^\beta}),(\frac{1}{2},\frac{1}{2^\beta})]\!] + [\![(\frac{1}{2},\frac{1}{2^\beta}),(1,1)]\!] .$$

The new current has the multiplicity 2 interval $2[\![(\frac{1}{2},-\frac{1}{2^\beta}),(\frac{1}{2},\frac{1}{2^\beta})]\!]$ which we may then replace by the sum

$$2[\![(\frac{1}{2},-\frac{1}{2^\beta}),(\frac{1}{3},-\frac{1}{3^\beta})]\!] + 2[\![(\frac{1}{3},-\frac{1}{3^\beta}),(\frac{1}{3},\frac{1}{3^\beta})]\!] + 2[\![(\frac{1}{3},\frac{1}{3^\beta}),(\frac{1}{2},-\frac{1}{2^\beta})]\!] .$$

Continuing we obtain a size-minimizing sequence whose mass approaches infinity. These do not converge as currents and the resulting formal countable sum of oriented intervals with integer multiplicities is *not* a current. We will see that this can be understood as a "scan". There are some positive results concerning size-minimization.

2.2 Theorem. [M] *If* $\operatorname{spt}(\partial T_0)$ *is a smooth* $m-1$*-dimensional submanifold of* $\mathbf{R}^{m+1}$ *that lies on the boundary of a smooth compact convex body, then there exists a size-minimizing rectifiable current* T *with* $\partial T = \partial T_0$.

The idea here is that one can modify a size-minimizing sequence using decompositions into oriented boundaries of sets which extend all the way to the boundary Γ of the convex body. Since $\Gamma \setminus \operatorname{spt} \partial T_0$ contains only finitely many components, one thus obtains a bound $\mathcal{H}^m$ almost everywhere on the densities in the sequence. So the masses are bounded, and one has convergence as currents. The lower semicontinuity of size under this condition was established in [A3]. In [DH1] we obtain, for a general codimension one rectifiable current T_0, (without the convex hull property of $\operatorname{spt}(\partial T_0)$) a weaker result concerning the existence of a *minimizing set.* An m-dimensional set $S \subset \mathbf{R}^n$ is *minimizing* [A2] relative to a compact set K if

$$\mathcal{H}^m[f(S)] \geq \mathcal{H}^m(S)$$

for any Lipschitz map $f : \mathbf{R}^n \to \mathbf{R}^n$ with $\overline{\{x : f(x) \neq x\}} \subset \mathbf{R}^n \setminus K$.

2.3 Theorem. [DH1] *If, in* $\mathbf{R}^{m+1}$, $T_0 \in \mathcal{R}_m$ *and* $\mathcal{H}^m(\operatorname{spt} \partial T_0)) = 0$, *then there exists a minimizing set* S *with respect to* $\operatorname{spt} \partial T_0$ *having* $\operatorname{spt} \partial T_0 \subset \overline{S}$.

Here is a brief outline of our construction. First we penalize the lack of compactness by choosing, for any $0 < \epsilon < 1$, a rectifiable current T_ϵ minimizing $\operatorname{Size}(T) + \epsilon \mathbf{M}(T)$ among rectifiable currents T with $\partial T_\epsilon = \partial T_0$. Second we observe that the (renormalized) measure

$$\mu_\epsilon = (\mathcal{H}^m \llcorner R_{T_\epsilon})(1 + \epsilon \theta_{T_\epsilon})$$

defines a (real) rectifiable varifold, *stationary* [Al] in $\mathbf{R}^{m+1} \setminus \operatorname{spt} \partial T_0$. That is, the first variation of the total mass of the varifold μ_ϵ vanishes for any deformation which fixes $\operatorname{spt} \partial T_0$. Since the total mass of each μ_ϵ is bounded by $\operatorname{Size}(T_0) + \mathbf{M}(T_0)$, we may by [Al], choose a sequence μ_{ϵ_j} weakly convergent to a rectifiable varifold μ. We show that the m density of this varifold is one at $\mathcal{H}^m$ almost every point of $\operatorname{spt} \mu \setminus \operatorname{spt} \partial T_0$. Thus $\mu = \mathcal{H}^m \llcorner S$ for some rectifiable set S.

Third, by modifying some arguments of Ambrosio, Fusco and Hutchinson [AKH], which proved the minimality of a limit of codimension one minimizing sets, we verify that S is the desired minimizing set. It is an interesting question to find conditions that will guarantee that the *currents* T_{ϵ_j} converge. While Morgan's example above shows that this will not happen in general, the condition that $\operatorname{spt} \partial T_0$ be a smooth submanifold may be sufficient. Even for the "soap-film case" of two-dimensional size-minimizers in $\mathbf{R}^3$, lack of a priori knowledge of the boundary behaviour prevents progress. In the next result, one avoids the boundary behaviour problem.

2.4 Theorem. [M], [DH1] *In a compact Riemannian 3-manifold, any two-dimensional homology class contains a size-minimizing rectifiable current.*

J. Taylor [T] classified the local interior structure of the minimizing set S. Up to a $\mathcal{C}^{1,\alpha}$ diffeomorphism of space, the neighborhood of an interior point is either a plane, three half-planes meeting at equal angles along a line, or 6 planar sectors meeting at equal dihedral angles at a point. This local structure and the compactness of S imply that there exists a global Lipschitz retraction ρ of some open neighborhood of S onto S. Since $\operatorname{spt} T_{\epsilon_j} = \operatorname{spt} \mu_{\epsilon_j}$ converge in Hausdorff metric to S, we find that, for some fixed, sufficiently large j, the retracted current $\rho_{\#} T_{\epsilon_j}$ is the desired size-minimizing homology representative. To use this retraction argument for the original Plateau boundary problem, one would need more information about the boundary behaviour of minimizing sets or (approximately) size-minimizing currents. One can also consider a *free boundary* or *obstacle* problem where the minimization occurs among rectifiable currents constrained to have their boundaries lie on a given hypersurface. Here, for two dimensions in three space, the local structure is easy to classify. Again up to a $\mathcal{C}^{1,\alpha}$ diffeomorphism of space, the neighborhood of a free boundary point is either a half-space perpendicular to the given hypersurface or else 3 quadrants at equal angles to each other, meeting the hypersurface orthogonally. Thus one obtains

2.5 Theorem. [DH2] *For any smoothly bounded region* $\Omega \subset \mathbf{R}^3$ *and any* $T_0 \in \mathcal{R}_2$ *with* $\operatorname{spt} T_0 \subset \mathbf{R}^3 \setminus \Omega$ *and* $\operatorname{spt} \partial T_0 \subset \partial\Omega$*, the relative homology class*

$$\{T \in \mathcal{R}_2 \ : \ \operatorname{spt} T \subset \mathbf{R}^3 \setminus \Omega, \ \operatorname{spt} \partial T_0 \subset \partial\Omega, \ \text{and} \\ \operatorname{spt}(T - T_0 - \partial S) \subset \partial\Omega \ \text{for some} \ S \in \mathcal{R}_3\}$$

contains a rectifiable current of least size.

Note that one can imagine Ω as the interior of a thick three-dimensional wire whose boundary surface is supporting the boundary of a soap film. Unfortunately this last existence theorem provides no obvious uniform mass bounds on the size-minimizing currents for a sequence of such thick wires $\overline{\Omega_i}$ shrinking to a one-dimensional smooth curve. There are many other functionals that may also give rise to some minimizing sequences which have unbounded mass and which have no subsequences convergent as currents. Perhaps the simplest are given by "fractional" powers of the density.

2.6 Definition. For any $\alpha \in [0,1]$ and $T \in \mathcal{R}_m$, the *α-mass* of T equals

$$\mathbf{M}_\alpha(T) = \int_{R_T} [\theta_T(x)]^\alpha \, d\mathcal{H}^m x .$$

The case $\alpha = 0$ corresponds to the Size$(T) = \mathcal{H}^m(R_T)$ and the case $\alpha = 1$ corresponds to the ordinary mass $\mathbf{M}(T)$. The special property of the range $0 \leq \alpha \leq 1$ is that the functional $\mathbf{M}_\alpha$ is weakly lower semicontinuous on mass bounded sequences. In case $m = 0$ one is only dealing with "collisions of atoms" and the result is elementary, based on the inequality

$$|i + j|^\alpha \leq |i|^\alpha + |j|^\alpha \text{ for any integers } i \text{ and } j ,$$

which is only valid for $\alpha \in [0,1]$. For larger m one may reduce to the case $m = 0$ by slicing which we now review.

3. Slicing

For any bounded Borel m form Ω on $\mathbf{R}^n$, let $T \llcorner \Omega$ denote the zero-dimensional current given by $(T \llcorner \Omega)(\psi) = T(\psi\Omega)$ for $\psi \in C_0^\infty(\mathbf{R}^n)$. From [F1], 4.3 we recall that if $f : \mathbf{R}^n \to \mathbf{R}^m$ is Lipschitz and $T \in \mathcal{R}_m$, then, for a.e. $y \in \mathbf{R}^m$ the slice $\langle T, f, y\rangle \in \mathcal{R}_0$ where

$$\langle T, f, y\rangle = \lim_{r \downarrow 0} T \llcorner f^\#\big(\frac{\chi_{\mathbf{B}_r(y)}}{\omega_m r^m}\Omega_m\big) \text{ with } \Omega_m = dx^1 \wedge \cdots \wedge dx^m .$$

In fact, for a.e. $y \in \mathbf{R}^m$, $R_T \cap f^{-1}\{y\}$ is a finite set of points x where the approximate tangent space of R_T exists with $f|R_T$ being approximately differentiable of rank m at x, and the slice is given by the formula

$$\langle T, f, y\rangle = \sum_{x \in R_T \cap f^{-1}\{y\}} \sigma(x)\theta_T(x)\delta_x$$

where $\sigma(x) = \operatorname{sgn}\big\langle \bigwedge_m Df(x)\vec{T}(x), \Omega_m\big\rangle$. One also has the integral formulas

$$\int_{\mathbf{R}^m} \langle T, f, y\rangle \, dy = T \llcorner f^\#\Omega_m ,$$

$$\int_{\mathbf{R}^m} \mathbf{M}_\alpha\langle T, f, y\rangle \, dy = \mathbf{M}_\alpha[T \llcorner f^\#\Omega_m] .$$

Recall also the compact space $\mathcal{P}$ of orthogonal projections of $\mathbf{R}^n$ onto $\mathbf{R}^m$, which has an invariant probability measure induced by the transitive action of $\mathbf{O}(n)$. In particular, for each increasing function $\lambda : \{1, \ldots, m\} \to \{1, \ldots, n\}$ we have the coordinate projection

$$p_\lambda : \mathbf{R}^n \to \mathbf{R}^m , \quad p_\lambda(x_1, \ldots, x_n) = (x_{\lambda(1)}, \ldots, x_{\lambda(m)}) .$$

By writing an m form $\phi \in \mathcal{D}(\mathbf{R}^n)$ in coordinates

$$\phi = \sum_\lambda \phi_\lambda dx^\lambda = \sum_\lambda \phi_\lambda p_\lambda^\# \Omega_m$$

with each $\phi_\lambda \in \mathcal{C}_0^\infty(\mathbf{R}^n)$, we see how the current T is expressed in terms of its zero-dimensional coordinate slices

$$\begin{aligned} T(\phi) &= \sum_\lambda T[\phi_\lambda p_\lambda^\# \Omega_m] = \sum_\lambda (T \llcorner p_\lambda^\# \Omega_m)(\phi_\lambda) \\ &= \sum_\lambda \int_{\mathbf{R}^m} \langle T, p_\lambda, y\rangle(\phi_\lambda)\, dy . \end{aligned}$$

Composing with rotations and integrating over $\mathbf{O}(n)$, we find a formula

$$\begin{aligned} T(\phi) &= (\operatorname{vol}\mathbf{O}(n))^{-1} \int_{\mathbf{O}(n)} \sum_\lambda \int_{\mathbf{R}^m} \langle T, p_\lambda \circ g, y\rangle(\phi \cdot (p_\lambda \circ g)^\# \Omega_m)\, dy\, dg \\ &= \binom{n}{m} (\operatorname{vol}\mathbf{O}(n))^{-1} \int_{\mathcal{P}} \int_{\mathbf{R}^m} \langle T, p, y\rangle(\phi \cdot p^\# \Omega_m)\, dy\, dp . \end{aligned}$$

One also has the integral geometric relation

$$\mathbf{M}_\alpha(T) = \beta(m,n) \int_{\mathcal{P}} \int_{\mathbf{R}^m} \mathbf{M}_\alpha \langle T, p, y\rangle\, dp\, dy .$$

This motivates us to define various classes of *scans* as measurable functions

$$\mathcal{T} : \mathcal{P} \times \mathbf{R}^m \to \{\,\text{zero} - \text{dimensional currents}\,\} .$$

In particular, a *rectifiable scan* is a measurable function

$$\mathcal{T} : \mathcal{P} \times \mathbf{R}^m \to \mathcal{R}_0$$

corresponding to some rectifiable set $R_\mathcal{T}$, some $\mathcal{H}^m$-measurable $\theta_\mathcal{T} : R_\mathcal{T} \to \{1, 2, \dots\}$ and some $\mathcal{H}^m$-measurable orientation $\vec{\mathcal{T}}$ of $R_\mathcal{T}$ so that, for almost all $(p, y) \in \mathcal{P} \times \mathbf{R}^m$,

$$\mathcal{T}(p,y) = \sum_{x \in R_\mathcal{T} \cap p^{-1}\{y\}} \sigma(x) \theta_\mathcal{T}(x) \delta_x$$

where $\sigma(x) = \operatorname{sgn} \langle \bigwedge_m p \vec{\mathcal{T}}(x), \Omega_m \rangle$. One defines

$$\mathbf{M}_\alpha(\mathcal{T}) = \beta(m,n) \int_{\mathcal{P}} \int_{\mathbf{R}^m} \mathbf{M}_\alpha(\mathcal{T}(p,y))\, dp\, dy ,$$

and sees that the scan $\mathcal{T}$ corresponds to a rectifiable current if and only if $\mathbf{M}_1(\mathcal{T}) < \infty$. For a rectifiable current $T \in \mathcal{R}_m$, an elementary Fourier transform argument shows that

$$\partial T = 0 \quad \text{if and only if} \quad \langle T, p, y\rangle(1) = 0 \ \text{ for almost all } (p,y) \in \mathcal{P} \times \mathbf{R}^m .$$

Thus, for 2 rectifiable scans $\mathcal{S}, \mathcal{T}$ we say

$$\partial \mathcal{S} = \partial \mathcal{T} \quad \text{if and only if} \quad (\mathcal{S} - \mathcal{T})(p,y)(1) = 0 \ \text{ for almost all } (p,y) .$$

3.1 Theorem. [DH1] *Suppose* $0 < \alpha \le 1$, $T_0 \in \mathcal{R}_m$, *and* $\mathcal{H}^m(\operatorname{spt} \partial T_0) = 0$. *Then there exists a rectifiable scan* $\mathcal{T}$ *with* $\partial \mathcal{T} = \partial T_0$ *and*

$$\mathbf{M}_\alpha(\mathcal{T}) = \inf\{\mathbf{M}_\alpha(T) : T \in \mathcal{R}_m,\ \partial T = \partial T_0\} .$$

The proof involves getting the convergence as scans of a subsequence of an $\mathbf{M}_\alpha$ minimizing sequence. We must work with a convergence that is weaker than the weak convergence of currents. To see this, consider the following

3.2 Example. Suppose $S_j = j^2\partial[\![\mathbf{B}_{1}j]\!]$ in $\mathbf{R}^2$. Thus S_j is an oriented circle of radius $\frac{1}{j}$ with multiplicity j^2. Then, in the weak topology of currents,

$$T_j \rightharpoonup \partial[\![\delta_{(0,0)}\mathbf{e}_1 \wedge \mathbf{e}_2]\!] \neq 0 ,$$

but, for $0 < \alpha < \frac{1}{2}$,

$$\mathbf{M}_\alpha(T_j) = j^{2\alpha} \cdot (\frac{2\pi}{j}) \to 0 .$$

An appropriate topology is the α flat distance [F12], [W] on $\mathcal{R}_m$

$$\mathcal{F}_\alpha(T_1, T_2) = \inf\{\mathbf{M}_\alpha(R) + \mathbf{M}_\alpha(S) \ : \ T_1 - T_2 = R + \partial S,\ R \in \mathcal{R}_m,\ S \in \mathcal{R}_{m+1}\} .$$

Recall now from [AK] that a measurable function f from $\mathbf{R}^m$ into a general metric space X has *finite total variation* if $\phi \circ f$ is BV for all Lipschitz $\phi : X \to \mathbf{R}$ and

$$\|Df\|(\mathbf{R}^m) \equiv \sup\{\int |D(\phi \circ f)| \ : \phi : X \to \mathbf{R},\ \operatorname{Lip}\phi \leq 1\} < \infty .$$

R. Jerrard [JS] observed that the slice of a normal current was of metric bounded variation with respect to the flat norm. The next result is the analogue for the $\mathcal{F}_\alpha$ distance.

3.3 Theorem. [DH1] *If $T \in \mathcal{R}_m$, $\partial T \in \mathcal{R}_{m-1}$, $\mathbf{M}_\alpha(T) + \mathbf{M}_\alpha(\partial T) < \infty$, and $p \in \mathcal{P}$, then*

$$f : \mathbf{R}^m \to (\mathcal{R}_0, \mathcal{F}_\alpha), \quad f(y) = \langle T, p, y\rangle$$

has finite total variation

$$\|Df\|(\mathbf{R}^m) \leq m\,[\mathbf{M}_\alpha(T) + \mathbf{M}_\alpha(\partial T)\,]$$

Proof. In case $m = 1$ the equation, for almost all $s < t$,

$$\langle T, p, s\rangle - \langle T, p, t\rangle = (\partial T) \llcorner p^{-1}[s,t] - \partial(T \llcorner p^{-1}[s,t])$$

implies that

$$\mathcal{F}_\alpha(\langle T, p, s\rangle, \langle T, p, t\rangle) \leq \mathbf{M}_\alpha(T \llcorner p^{-1}[s,t]) + \mathbf{M}_\alpha((\partial T) \llcorner p^{-1}[s,t]) .$$

Summing over almost all partitions of $\mathbf{R}$ then gives that the essential variation

$$\operatorname{essvar}(f) \leq \mathbf{M}_\alpha(T) + \mathbf{M}_\alpha(\partial T) ,$$

which implies the case $m = 1$. For $m = 2, 3, \ldots$, we use the formula [F1], 4.5.9.27

$$\int_{\mathbf{R}^m} |D\psi|$$

$$\leq \sum_{i=1}^{m} \int_{\mathbf{R}^{m-1}} \operatorname{essvar} \psi(y_1, \ldots, y_{i-1}, \cdot, y_{i+1}, \ldots, y_m)\, dy_1 \ldots dy_{i-1} dy_{i+1} \ldots dy_m$$

with $\psi = \phi \circ f$ to get the desired bound. This estimate gives the desired variation in y. To get a similar variation bound in (p, y) one may use, for almost all (p, y), the formula

$$\langle T, p, y\rangle \;=\; \Pi_{\#}\langle T \times [\![\mathcal{P}]\!], P, (y, p)\rangle$$

where $\Pi(x, q) = x$ and $P(x, q) = \big(q(x), q\big)$ for $(x, q) \in \mathbf{R}^n \times \mathcal{P}$. From the usual BV compactness for real-valued functions it is not difficult to derive the following metric-space-valued version:

3.4 Theorem. [DH1] *Suppose N is a Riemannian manifold, Y is a separable metric space, $M : Y \to \mathbf{R}^+$ is lower semicontinuous, and $M^{-1}[0, R]$ is sequentially compact for all $R > 0$. If $f_j : N \to Y$ is measurable with*

$$\Lambda \;=\; \sup_j \|Df_j\|(N) \;+\; \int_N M\big(f_j(x)\big) \;<\; \infty\;,$$

then some subsequence $f_{j'}$ converges pointwise a.e. to a measurable $f : N \to Y$ with

$$\|Df\|(N) \;+\; \int_N M\big(f(x)\big) \;\le\; \Lambda\;.$$

To prove the existence of a suitable scan $\mathcal{T}$ for Theorem 3.1, we can now apply Theorem 3.3, the remark after, and Theorem 3.4 with

$$N = \mathcal{P} \times \mathbf{R}^m,\; Y = (\mathcal{R}_0, \mathcal{F}_\alpha),\; M = \mathbf{M}_\alpha,\; f_j(p, y) = \langle T_j, p, y\rangle.$$

The lower semicontinuity follows from Fatou's lemma and the lower semicontinuity of $\mathbf{M}_a$ on $(\mathcal{R}_0, \mathcal{F}_\alpha)$. To verify the rectifiability of the limiting scan, we obtain a rectifiable varifold as in the proof of Theorem 2.3 and then show that our scan is necessarily concentrated on an $\mathcal{H}^m$-finite rectifiable concentration set of this varifold. Almost every projection is transverse a.e. to this rectifiable set and one works with the convergence of the slices in these directions to eventually get the desired consistently defined multiplicity $\theta_{\mathcal{T}}$ and orientation $\vec{\mathcal{T}}$. In [DH2] we obtain a general compactness theorem in the class of rectifiable scans having rectifiable boundaries and having $\mathbf{M}_\alpha + \mathbf{M}_a\partial$ uniformly bounded. This gives an existence theory for various Plateau problems. We also obtain the optimal interior partial regularity estimate for $\mathbf{M}_\alpha$ minimizers:

3.5 Theorem. [DH2] *For any m-dimensional $\mathbf{M}_\alpha$ minimizing rectifiable scan $\mathcal{T}$ in $\mathbf{R}^n$ and $\epsilon > 0$*

$$\mathcal{H}^{m-1+\epsilon}\big(\mathrm{Sing}\,(\mathrm{spt}\,\mathcal{T} \setminus \mathrm{spt}\,\partial\mathcal{T})\big) \;=\; 0\;.$$

Moreover, $\mathcal{T} \llcorner K$ has finite mass for every compact $K \subset \mathbf{R}^n \setminus \mathrm{spt}\,\partial\mathcal{T}$.

The proof uses the following (roughly stated)

3.6 Lemma. *If $0 < \alpha < 1$, then, near any point a of $\mathrm{spt}\,\mathcal{T} \setminus \mathrm{spt}\,\partial\mathcal{T}$ where $\mathcal{T}$ has a multiplicity ν tangent plane, the scan $\mathcal{T}$ is mostly a multiplicity ν graph of a <u>single</u> function.*

In shrinking cylinders about such a point, the ν excess [F1], 5.3, as well as the ordinary excess, approaches zero. This is in sharp contrast to the case $\alpha = 1$ of mass-minimizing currents. For example, the complex cusp

$$\{(z,w) \in \mathbf{C} \times \mathbf{C} \; : w^2 = z^3\}$$

supports a multiplicity one mass-minimizing rectifiable current with a multiplicity two tangent plane at $(0,0)$. The proof of Lemma 3.6 is based on a multi-valued graphical approximation followed by a squashed comparison current.

4. Another fractional integrand

In the works [HR1], [HR2] treating various energy-bounded sequences of Sobolev mappings, one encounters rectifiable currents with bounds on the integral of fractional powers of the (ordinary) mass of slice. In our notations, the analogous situation is to consider for $T \in \mathcal{R}_m$ and $0 < \alpha \leq 1$, the integral

$$\tilde{\mathbf{M}}_\alpha(T) \;=\; \beta(m,n) \int_{\mathcal{P}} \int_{\mathbf{R}^m} \big(\mathbf{M}\langle T,p,y\rangle\big)^\alpha dp\, dy \;.$$

Note the inequalities

$$\tilde{\mathbf{M}}_\alpha(T) \;\leq\; \mathbf{M}_\alpha(T), \;\; \mathbf{M}(T) \leq \tilde{\mathbf{M}}_1(T) \leq \mathbf{M}(T) \;.$$

4.1 Example. Consider in $\mathbf{R}^2$ the concentric, multiplicity one circles

$$T_j = \sum_{i=1}^{j} \partial[\![\mathbf{B}_{1/i}]\!] \in \mathcal{R}_1 .$$

Then, as $j \to \infty$,

$$\mathbf{M}_\alpha(T_j) \;=\; \mathbf{M}(T_j) \to \infty \;, \text{ but } \; \sup_j \tilde{\mathbf{M}}_\alpha(T_j) \;<\; \infty \;.$$

Here the limit is a rectifiable scan with $\tilde{\mathbf{M}}_\alpha$ finite, but with a concentration set of infinite Hausdorff measure. As in [HR1], [HR2], one may again work with $(\mathcal{R}_0, \mathcal{F}_\alpha)$ to obtain the lower semicontinuity of $\tilde{\mathbf{M}}_\alpha$. One can obtain the existence of $\tilde{\mathbf{M}}_\alpha$ minimizing scans by using a weak-type bound in the Lorentz space $L^{\frac{1}{\alpha},\infty}$. See [HR1], Section 9. However the partial regularity or even rectifiability of $\tilde{\mathbf{M}}_\alpha$ minimizers is unknown. See [HR1], Section 8.2.

5. Related problems

In the study of $\mathbf{M}_a$ minimizers, one may replace the power function θ^α by a smooth concave unbounded increasing function $H(\theta)$ with $H(0) = 0$ and $H(1) = 1$. Use of scans accommodates as well treatment of the case when $H = H(\theta, x)$ is also allowed to depend smoothly on the space variable x. By using $H(\theta)$ as an alternate norm on the group of integers, a rectifiable minimizer may also be found in a generalized class of flat chains following the works of Fleming [F2] and White [W].

The close relation between rectifiability and slicing explained in [AK] and [W] was an important motivation for the definition of scans in [HR1] and [DH1]. One-dimensional *flat* $\mathbf{M}_\alpha$ minimizers are also applied to describe transport paths in [X1], [X2]. The $\mathbf{M}_\alpha$ functional reflects the efficiency of combining paths in various distribution systems, such as mail delivery, the circulatory system, etc. Here one obtains, for any two probability measures μ_0, μ_1 in $\mathbf{R}^n$, a *flat* 1 chain T which minimizes $\mathbf{M}_\alpha$ subject to the constraint $\partial T = \mu_1 - \mu_0$ as zero-dimensional currents. In [X1], the transport path T has positive *real* density function θ_T with values in $(0,1]$ because $\mathbf{M}(\mu_1) = 1 = \mathbf{M}(\mu_2)$. So one here has the inequality

$$\mathbf{M}_\alpha(T) \geq \mathbf{M}(T) ,$$

which is just the opposite of the inequality that we had in our study [DH1], [DH2] of integer-multiplicity rectifiable currents. In contrast to [DH1], [DH2], $\mathbf{M}_\alpha$ minimization for transport paths always gives finite mass *currents*. In [X2], Q. Xia proves the precise local interior regularity of an $\mathbf{M}_\alpha$ minimizing path: that, in $\mathbf{R}^n \setminus (\operatorname{spt}\mu_1 \cup \operatorname{spt}\mu_2)$, T is locally a finite collection of oriented intervals with multiplicities. Finally we have begun work in [DH3] on carrying over various results of geometric measure theory to rectifiable scans. There we introduce the notion of a rectifiable scan in a metric space X, and give some results generalizing the work of L.Ambrosio and B. Kirchheim [AK] on currents in metric spaces. The idea is that an m-dimensional rectifiable scan in X is a measurable function

$$\mathcal{T} \;:\; \operatorname{Lip}(X, \mathbf{R}^m) \times \mathbf{R}^m \;\rightarrow\; \mathcal{R}_0(X)$$

which admits a representation in terms of an m-dimensional rectifiable [AK] set $R_\mathcal{T} \subset X$, and integer density function $\theta_\mathcal{T}$, and an orientation $\vec{\mathcal{T}}$ of $R_\mathcal{T}$. (The notion of orientation requires some effort to describe.) For a Lipschitz map $g : X \rightarrow Y$ of metric spaces and m-dimensional rectifiable scan $\mathcal{T}$ on X, the push-forward $g_\#\mathcal{T}$ defined by

$$(g_\#\mathcal{T})(f,y) \;=\; g_\#\big(\mathcal{T}(f \circ g, y)\big)$$

for almost all $(f,y) \in \operatorname{Lip}(X,\mathbf{R}^m) \times \mathbf{R}^m$, is a rectifiable scan on Y. An $m-1$-dimensional scan $\mathcal{S}$ is the *boundary*, $\partial\mathcal{T}$, of $\mathcal{T}$ if, for all points $a \in X$,

$$\mathcal{S}(g,z) \;=\; \lim_{r\downarrow 0} \mathcal{T}\big((g, \operatorname{dist}(\cdot, a)), (r,z)\big)$$

for almost all $(g,z) \in \operatorname{Lip}(X, \mathbf{R}^{m-1}) \times \mathbf{R}^{m-1}$. One may again impose bounds on $\mathbf{M}_\alpha$ and $\mathbf{M}_\alpha\partial$ along with suitable topological bounds on supports to obtain scan compactness theorems.

References

[Al] W.K. Allard, *On the first variation of a varifold.* Ann. of Math.(2) **95** (1972), 417–491.

[A1] F.J. Almgren, *Some interior regularity theorems for minimal surfaces and an extension of Bernstein's theorem.* Ann. Math. **84**(1966), pp. 277–292.

[A2] F.J.Almgren, *Existence and regularity almost everywhere of solutions to elliptic variational problems with constraints.* Mem. Amer. Math. Soc. **4**(1976), no. 165.

[A3] F.J. Almgren, *Deformations and multi-valued functions* in *Geometric measure theory and the calculus of variations.* AMS Proc. Sym. Pure Math **44**(1996), pp. 29–130.

[A4] F.J. Almgren, *Almgren's big regularity paper* World Scientific Publishing Co., Inc., River Edge, NJ, 2000.

[AFK] L. Ambrosio, N. Fusco, and J. Hutchinson, *Higher integrability of the gradient and dimension of the singular set for minimizers of the Mumford-Shah functional.* Calculus of Variations and P.D.E. **16**(2003)(2), 187–215.

[AK] L. Ambrosio and B. Kirchheim, *Currents in metric spaces.* Acta Math. **185**(2000), no.1, 1–80.

[BDG] E. Bombieri, E. De Giorgi, and E. Giusti, *Minimal cones and the Bernstein problem* Invent. Math. **7**(1969), 243–268.

[C] S. Chang, *Two-dimensional area minimizing integral currents are classical minimal surfaces.* J.A.M.S. **1**(1988), pp. 699–778.

[D] E. De Giorgi, *Frontiere orientate di misura minima.* Sem. Mat. Scuola Norm. Sup. Pisa, 1961.

[DH1] T. De Pauw and R. Hardt, *Size minimization and approximating problems.* Calculus of Variations and P.D.E. **17**(2003)(4), 405–442.

[DH2] T. De Pauw and R. Hardt, *Partial regularity of scans minimizing a fractional density mass.* In preparation.

[DH3] T. De Pauw and R. Hardt, *Rectifiable scans in a metric space.* In preparation.

[F1] H. Federer, *Geometric measure theory.* Springer-Verlag, Berlin and New York, 1969.

[F1] H. Federer, *The singular sets of area minimizing rectifiable currents with codimension one and of area minimizing flat chains modulo two with arbitrary codimension.* Bull. A.M.S. **76**(1970), pp. 767–771.

[FF] H. Federer and W. Fleming, *Normal and integral currents.* Ann. Math. **72**(1960), pp. 458–520.

[Fl1] W. Fleming, *On the oriented Plateau problem*, Rend. Circ. Mat. Palermo, (2) **11**(1962), pp.1–22.

[Fl2] W. Fleming, *Flat chains over a coefficient group.* Trans.A.M.S. **121**(1966), 160–186.

[HR1] R. Hardt and T. Rivière, *Connecting topological Hopf singularities.* To appear in Annali Sc. Norm. Sup. Pisa.

[HR2] R. Hardt and T. Rivière, *Connecting rational homotopy type singularities.* In preparation.

[HS] R. Hardt and L. Simon, *Boundary regularity and embedded solutions for the oriented Plateau problem.* Ann.Math.(2) **110**(1979), pp. 439–486.

[JS] R. Jerrard and M. Soner, *Functions of bounded higher variation*, Ind. U. Math. J. **51**(2002)(3), 645–677.

[M] F. Morgan, *Size-minimizing rectifiable currents*, Invent. Math. **96**(2)(1989), pp. 333–348.

[S] L. Simon, *Lectures on geometric measure theory.* Proc. Centre for Math. Anal. **3** (1983) Australian National University, Canberra.

[T] J.E. Taylor, *The structure of singularities in soap-bubble-like and soap-film-like minimal surfaces.* Ann. Math.(2) **103**(1976), pp. 489–539.

[Tr] D. Triscari, *Sulle singolarità delle frontiere orientate di misura minima nello spazio euclideo a 4 dimensioni.* Matematiche (Catania) **18**(1963), pp. 139–163.

[Ss] J. Simons, *Minimal varieties in Riemannian manifolds.* Ann. Math. **88**(1968), pp. 62–105.

[W] B. White *Rectifiability of flat chains.* Ann. of Math.(2) **150**(1999), no.1, pp. 165–184.

[X1] Q. Xia, *Optimal paths related to transport problems.* Commun. Contemp. Math. **5**(2003), pp. 251–279.

[X2] Q. Xia, *Interior regularity of optimal transport paths.* Preprint 2002.

Thierry De Pauw[1]
Université Catholique de Louvain
Belgium
e-mail: `depauw@math.ucl.ac.be`

Robert Hardt[2]
Rice University
USA
e-mail: `hardt@rice.edu`

[1]Chercheur qualifié of the FNRS, Belgium and partially supported a Marie Curie fellowship from the European Community contract HMPF-CT-2001-01235.
[2]Research partially supported by NSF grant DMS-0306294.

Progress in Nonlinear Differential Equations
and Their Applications, Vol. 59, 33–42

Bubbling of Almost-harmonic Maps between 2-spheres at Points of Zero Energy Density

Peter Topping

Abstract. We show that bubbling of almost-harmonic maps between 2-spheres has very different behaviour depending on whether or not bubbles develop at points in the domain at which the energy density of the body map is zero. We also see that this translates into different behaviour for the harmonic map flow. In [11] we obtained results, assuming nonzero bubble point density for certain bubbles, forcing the harmonic map flow to converge uniformly and exponentially to its limit. This involved proving a type of nondegeneracy for the harmonic map energy (a 'quantization' estimate') and an estimate on certain bubble scales (a 'repulsion' estimate). Here we show that without the nonzero bubble point density hypothesis, both the quantization and repulsion estimates fail, and we construct a flow in which the convergence is no longer exponentially fast.

1. Introduction

In this article, we study the harmonic map energy functional

$$E(u) := \frac{1}{2}\int_{S^2} |\nabla u|^2$$

for maps $u : S^2 \to S^2 \hookrightarrow \mathbb{R}^3$. We are interested in the behaviour of E near its critical points – known as harmonic maps – and near more general objects – known as bubble trees – which can be thought of as critical points at infinity. Such information is related to the behaviour of the L^2-gradient flow for E – known as the harmonic map flow – which is a solution of

$$\frac{\partial u}{\partial t} = \mathcal{T}(u(t)), \tag{1.1}$$

where the tension $\mathcal{T}(u) \in \Gamma(u^*(TS^2))$ is defined to be the negation of the L^2-gradient of E. (See [11] or [4] for more details in this case.) The flow, when smooth, dissipates energy according to

$$\frac{d}{dt}E(u(t)) = -\|\mathcal{T}(u(t))\|^2_{L^2(S^2)},$$

which integrates to the useful identity

$$\int_s^\infty \|\mathcal{T}(u(t))\|^2_{L^2(S^2)} dt = E(u(s)) - \lim_{t\to\infty} E(u(t)) < \infty. \tag{1.2}$$

Harmonic maps between 2-spheres are well known to be precisely the rational maps and their complex conjugates (see [4, (11.5)]). In particular, being conformal maps from a surface, their energy is precisely the area of their image, and thus

$$E(u) = 4\pi|\deg(u)| \in 4\pi\mathbb{Z},$$

for any harmonic u. Since the harmonic maps occur in families, the energy E is degenerate near each critical point. However, by combining work of Gulliver-White [5] and L. Simon [7], [8, Lecture 3, Section 2] it is possible to prove a substitute for nondegeneracy, that

$$0 \leq E(u) - E(h) \leq C\|\mathcal{T}(u)\|^2_{L^2(S^2)}, \tag{1.3}$$

for u sufficiently close[1] to a harmonic map $h : S^2 \to S^2$. Estimate (1.3) can be used to analyse the asymptotics of the harmonic map heat flow when it converges smoothly (without bubbling as described below) at infinite time. In particular, for a flow converging smoothly to a limit u_∞ as $t \to \infty$,

$$\frac{\partial}{\partial t}\left[E(u(t)) - E(u_\infty)\right] = -\|\mathcal{T}(u(t))\|^2_{L^2(S^2)} \leq -\frac{1}{C}\left[E(u(t)) - E(u_\infty)\right],$$

for sufficiently large t, and we see that $E(u(t)) - E(u_\infty)$ must decay to zero exponentially fast. Moreover, (1.3) allows us to calculate

$$\begin{aligned} -\frac{d}{dt}\left[E(u(t)) - E(u_\infty)\right]^{\frac{1}{2}} &= \frac{1}{2}\left[E(u(t)) - E(u_\infty)\right]^{-\frac{1}{2}} \|\mathcal{T}(u(t))\|^2_{L^2(S^2)} \\ &\geq \frac{1}{2\sqrt{C}}\|\mathcal{T}(u(t))\|_{L^2(S^2)}, \end{aligned}$$

and thus, for $s \in [0,\infty)$ (and a new constant C independent of s)

$$\int_s^\infty \|\mathcal{T}(u(t))\|_{L^2(S^2)} dt \leq C\left[E(u(s)) - E(u_\infty)\right]^{\frac{1}{2}}. \tag{1.4}$$

To begin with, this implies the finiteness of the left-hand side of (1.4) (which is simply false for the harmonic map flow in general, despite (1.2)). In addition, the exponential energy decay then forces the left-hand side of (1.4) to decay to zero exponentially fast. By returning to the equation (1.1) we then must have exponential convergence of $u(t)$ to u_∞ as $t \to \infty$ in $L^2(S^2)$ and thus in C^k – for any k – by interpolation. (Even the convergence of $u(t)$ to u_∞ in $L^2(S^2)$ *uniformly* as $t \to \infty$ is false in general for the harmonic map flow – see [10].) Unfortunately, in general the harmonic map flow does not converge smoothly at infinite time (see [1] or the examples below). The best we can hope for is 'bubbling' convergence as the flow approaches a 'bubble tree' which can be thought of as a critical point at

[1] for example, close in C^3

infinity for E. We first describe bubbling in the context of sequences of almost-harmonic maps. The following theorem is a combination of the seminal work of Struwe [9] and the improvements of Qing [6], Ding-Tian [3], and Wang [12]. See [11, Theorem 1.1] for a more sophisticated version.

Theorem 1.1. *Suppose that $u_n : S^2 \to S^2 \hookrightarrow \mathbb{R}^3$ is a sequence of smooth maps which satisfy $E(u_n) < M$ for some constant M, and all $n \in \mathbb{N}$, and $\mathcal{T}(u_n) \to 0$ in $L^2(S^2)$ as $n \to \infty$. Then we may pass to a subsequence in n, and find a harmonic map $u_\infty : S^2 \to S^2$, and a set $\{x^1, \ldots, x^m\} \subset S^2$ such that*

(a) *$u_n \rightharpoonup u_\infty$ weakly in $W^{1,2}(S^2)$,*

(b) *$u_n \to u_\infty$ strongly in $W^{2,2}_{loc}(S^2 \backslash \{x^1, \ldots, x^m\})$.*

Moreover, for each x^j, if we precompose each u_n and u_∞ with an inverse stereographic projection sending $0 \in \mathbb{R}^2$ to $x^j \in S^2$ (and continue to denote these compositions by u_n and u_∞ respectively) then for $i \in \{1, \ldots, k\}$ (for some $k \in \mathbb{N}$ dependent on j) there exist sequences $a_n^i \to 0 \in \mathbb{R}^2$ and $\lambda_n^i \downarrow 0$ as $n \to \infty$, and nonconstant harmonic maps $\omega^i : S^2 \to S^2$ (which we precompose with the same inverse stereographic projection to view them also as maps $\mathbb{R}^2 \cup \{\infty\} \to S^2$) such that:

(i)
$$\frac{\lambda_n^i}{\lambda_n^j} + \frac{\lambda_n^j}{\lambda_n^i} + \frac{|a_n^i - a_n^j|^2}{\lambda_n^i \lambda_n^j} \to \infty,$$

as $n \to \infty$, for each unequal $i, j \in \{1, \ldots, k\}$.

(ii)
$$\lim_{\mu \downarrow 0} \lim_{n \to \infty} E(u_n, D_\mu) = \sum_{i=1}^{k} E(\omega^i).$$

(iii) *For each $i \in \{1, \ldots, k\}$ there exists a finite set of points $\mathcal{S} \subset \mathbb{R}^2$ (which may be empty, but could contain up to $k-1$ points) with the property that*
$$u_n(a_n^i + \lambda_n^i x) \to \omega^i(x),$$
in $W^{2,2}_{loc}(\mathbb{R}^2 \backslash \mathcal{S})$ as $n \to \infty$.

We refer to the map $u_\infty : S^2 \to S^2$ as a 'body' map, and the maps $\omega^i : S^2 \to S^2$ as 'bubble' maps, or simply 'bubbles.' The points $\{x^1, \ldots, x^m\}$ will be called 'bubble points,' and the λ_n^i 'bubble scales.' Since each ω^i is a nonconstant harmonic map between 2-spheres, the energy of each must be at least 4π. In anticipation of an application of Theorem (1.1) to the harmonic map flow, we now seek a generalisation of (1.3); we want to understand the behaviour of E for maps not necessarily C^k-close to a harmonic map, but with small tension in L^2. (Such maps need not be even $W^{1,2}$-close to a harmonic map – see [11].) In [11, Theorem 1.2] we proved such a generalisation, building on our earlier work in [10]. Here we state a weaker version with simpler hypotheses.

Theorem 1.2. *Suppose that we have a sequence of maps $u_n : S^2 \to S^2$ satisfying the hypotheses of Theorem 1.1, and that we pass to a subsequence and find a limit u_∞, bubble points $\{x^j\}$ and bubble data ω^i, λ_n^i, a_n^i at each bubble point – as we* *know we can from Theorem* 1.1. *Suppose that at each bubble point, only one bubble develops – i.e., that in the language of Theorem* 1.1, $k = 1$ *at each* x^j. *Suppose further that the energy density of the body map is nonzero at each bubble point – i.e., that* $|\nabla u_\infty|(x^j) \neq 0$ *for all* $j \in \{1, \ldots, m\}$. *Then there exist constants* $C > 0$ *and* $k \in \mathbb{N} \cup \{0\}$ *such that after passing to a subsequence, the energy is quantized according to*

$$|E(u_n) - 4\pi k| \leq C \|\mathcal{T}(u_n)\|^2_{L^2(S^2)}. \tag{1.5}$$

A further part of [11, Theorem 1.2] is a 'repulsion' estimate which controls *some* of the bubble scales λ_n^i by:

$$\lambda_n^i \leq \exp\left[-\frac{1}{C\|\mathcal{T}(u_n)\|^2_{L^2(S^2)}}\right]. \tag{1.6}$$

Remark 1.3. The full version of [11, Theorem 1.2] allows multiple bubbling at each point subject to a compatibility condition, and only requires the body map to have zero energy density at certain bubble points. It is at *these* bubble points that the bubble scales satisfy the repulsion estimate (1.6).

One of the results of this paper is that both the quantization estimate (1.5) and the repulsion estimate (1.6) may fail if we allow arbitrary bubbles to occur at points where the energy density of u_∞ is zero. See Corollary 2.2 below. Theorem 1.1 gives us a basic understanding of the asymptotics of the harmonic map flow, as we now describe. By considering (1.2), given any sufficiently regular harmonic map flow u, we may find a sequence of times $t_n \to \infty$ such that $\mathcal{T}(u(t_n)) \to 0$ in $L^2(S^2)$. We may then apply Theorem 1.1 to the sequence of maps $\{u(t_n)\}$ to get bubbling convergence. (So far, this says nothing about the behaviour of the flow at times between t_n and t_{n+1}.) It turns out to be possible (see [11] for details) to apply Theorem 1.2 to flows with appropriate bubbling, in order to obtain a quantization estimate for $u(t)$ for *all* sufficiently large t – not just for the maps $u(t_n)$. We state this result for smooth flows – note that the standard Struwe flow (as in [9]) is smooth for sufficiently large time, and we are only concerned here with the asymptotics of the flow. As before, the results in [11] are somewhat stronger.

Theorem 1.4. *Suppose that* $u : S^2 \times [0, \infty) \to S^2$ *is a smooth solution of* (1.1). *Then after picking times* $t_n \to \infty$ *so that* $\mathcal{T}(u(t_n)) \to 0$ *in* $L^2(S^2)$ *and analysing the sequence* $\{u(t_n)\}$ *with Theorem* 1.1, *if only one bubble develops at each bubble point and the energy density of the body map* u_∞ *is nonzero at each bubble point, then there exists* $C > 0$ *independent of* t *such that*

$$0 \leq E(u(t)) - \overline{E} \leq C\|\mathcal{T}(u(t))\|^2_{L^2(S^2)}, \tag{1.7}$$

for sufficiently large t, *where*

$$\overline{E} := \lim_{t\to\infty} E(u(t)) \in 4\pi\mathbb{Z}.$$

Remark 1.5. Exactly as described earlier for smoothly converging flows, we can use an estimate such as (1.7) to deduce exponential decay of both $\left[E(u(t)) - \overline{E}\right]$

and $\int_t^\infty \|\mathcal{T}(u(s))\|_{L^2(S^2)}ds$. This time, the consequence is the following corollary (cf. [11, Theorem 1.7]).

Corollary 1.6. *Under the hypotheses of Theorem 1.4, for any $k \in \mathbb{N}$ and $\Omega \subset\subset S^2\backslash\{x^1, \ldots, x^m\}$ (where $\{x^1, \ldots, x^m\}$ are the bubble points) we have exponential decay of*

$$\|u(t) - u_\infty\|_{L^2(S^2)} + \|u(t) - u_\infty\|_{C^k(\Omega)}, \tag{1.8}$$

as $t \to \infty$.

A priori, not even uniform convergence $u(t) \to u_\infty$ in $L^2(S^2)$ as $t \to \infty$ is clear. In fact, for more elaborate target manifolds than S^2, this may fail – see [10]. A further result of this paper is that if we allow a smooth flow to develop bubbles at points in the domain where the energy density of u_∞ is zero, then the convergence of the flow need not be exponential – see Theorem 2.3 below. The combination of this result and [11] leads to apparently the first situation for the harmonic map flow in which the *position* of bubble points in the domain is shown to have a bearing on the properties of the bubbling.

2. Details of results

We first need the results of some energy and tension computations for a simple map between 2-spheres.

Proposition 2.1. *Given $d \in \mathbb{N}$ and $\varepsilon \in (0, 1]$, consider the map $w_\varepsilon : S^2 \to S^2$ defined in terms of stereographic complex coordinates by*

$$w_\varepsilon(z) = \varepsilon^d(z^d - \bar{z}^{-d}).$$

Then

(a)
$$c_0 d\varepsilon^{2d} \leq E(w_\varepsilon) - 8\pi d \leq C_0 d\varepsilon^{2d}, \tag{2.1}$$
for universal $c_0, C_0 > 0$;

(b)
$$c_1\varepsilon^{2d-1} \leq \|\mathcal{T}(w_\varepsilon)\|_{L^2(S^2)} \leq C_1\varepsilon^{2d-1}, \tag{2.2}$$
for $c_1, C_1 > 0$ dependent only on d.

Given these calculations (see Section 3 for comments on the details) we may define a sequence of maps $u_n : S^2 \to S^2$ by $u_n := w_{1/n}$ to establish the following corollary.

Corollary 2.2. *For any $d \in \mathbb{N}$, there exist $k \in \mathbb{N}$ and a sequence of smooth maps $u_n : S^2 \to S^2$ with $\mathcal{T}(u_n) \to 0$ in $L^2(S^2)$, such that*

$$c\|\mathcal{T}(u_n)\|_{L^2(S^2)}^{\frac{2d}{2d-1}} \leq |E(u_n) - 4\pi k| \leq C\|\mathcal{T}(u_n)\|_{L^2(S^2)}^{\frac{2d}{2d-1}},$$

with $c, C > 0$ dependent only on d. In particular, the quantization estimate (1.5) fails for the sequence $\{u_n\}$.

Moreover, if we apply Theorem 1.1 *to the sequence* $\{u_n\}$, *then the resulting bubble scales must satisfy*

$$c\|\mathcal{T}(u_n)\|_{L^2(S^2)}^{\frac{1}{2d-1}} \leq \lambda_n \leq C\|\mathcal{T}(u_n)\|_{L^2(S^2)}^{\frac{1}{2d-1}},$$

for some $c, C > 0$ *independent of* n.

The second part of this corollary should be compared to the repulsion estimate in [11] mentioned in (1.6). The sequence u_n converges to a constant body map with bubbles of degree d and $-d$. An alternative example, with slightly different exponents, would arise by defining instead $u_n(z) = w_{1/n}(nz)$. In this case we would get a degree d body map with a single degree $-d$ bubble. Next we state precisely our result for the harmonic map flow.

Theorem 2.3. *Fix an integer* $d > 1$ *and define* $u_0 : S^2 \to S^2$ *by*

$$u_0(z) = z^d - \bar{z}^{-d}.$$

Then there exist $\eta = \eta(d) > 0$ *and a unique smooth solution* $u : S^2 \times [0, \infty) \to S^2$ *of the harmonic map heat equation* (1.1) *with* $u(0) = u_0$, *and for* $t \geq 1$ *the solution has the following properties:*

(a) $$\|u(t) - u_\infty\|_{L^2(S^2)} \geq \eta t^{-\frac{1}{2(d-1)}}; \tag{2.3}$$

(b) $$\|\mathcal{T}(u(t))\|_{L^2(S^2)} \geq \eta t^{-\frac{d}{d-1}}; \tag{2.4}$$

(c) $$\int_t^\infty \|\mathcal{T}(u(s))\|_{L^2(S^2)} ds \geq \eta t^{-\frac{1}{d-1}}; \tag{2.5}$$

(d) $$E(u(t)) - \lim_{s\to\infty} E(u(s)) \geq \eta t^{-\frac{d+1}{d-1}}; \tag{2.6}$$

Inequality (2.3) implies that $u(t)$ does not converge exponentially to u_∞, and should be compared to (1.8). (Note that for the flow in question, $u(t)$ does nevertheless converge to u_∞ as $t \to \infty$.) Inequalities (2.5) and (2.6) imply in particular that their left-hand sides cannot decay exponentially to zero as $t \to \infty$, and should be compared to Remark 1.5.

3. Elements of the calculations and proofs

Although it is the complex coordinates which help us to pick natural maps for this problem, we are better off calculating in spherical polar coordinates (θ, ϕ) in order to later apply the comparison principle and exploit rotational symmetry. With our convention, $\theta \in [0, \pi]$ represents the 'latitudinal' angle from the *vertical*, and ϕ measures longitude. We will be considering, for integral $d \geq 1$, maps of the form

$$(\theta, \phi) \to (\beta(\theta), d\,\phi), \tag{3.1}$$

for appropriate $\beta : [0, \pi] \to \mathbb{R}$. The energy of such a map is given by

$$E = \pi \int_0^\pi \left[(\beta'(\theta))^2 \sin\theta + \frac{d^2}{\sin\theta} \sin^2\beta(\theta)\right] d\theta, \tag{3.2}$$

and the tension may be written

$$\mathcal{T} = \tau(\beta)\frac{\partial}{\partial\beta}$$

where $\tau(\beta) : [0,\pi] \to \mathbb{R}$ may be calculated[2] to be

$$\tau(\beta) = \beta''(\theta) + \frac{1}{\tan\theta}\beta'(\theta) - \frac{d^2 \sin\beta(\theta)\cos\beta(\theta)}{\sin^2\theta}, \tag{3.3}$$

and we have abused notation by using β both as a function and as a target coordinate.

Proof. (Proposition 2.1.) The map w_ε of Proposition 2.1 is of the form (3.1) with $\beta \equiv \alpha$ where

$$\alpha(\theta) = \alpha_\varepsilon(\theta) := 2\tan^{-1}\left[\varepsilon^d\left(\tan^d\frac{\theta}{2} - \tan^{-d}\frac{\theta}{2}\right)\right]. \tag{3.4}$$

Plugging this into (3.2) and using the substitution $t = \tan\frac{\theta}{2}$, we eventually find that

$$E - 8\pi d = 32\pi d^2\varepsilon^{2d}\int_0^1 \frac{t^{2d-1}}{(1+\varepsilon^{2d}(t^d - t^{-d})^2)^2}dt,$$

which then gives (2.1) of the proposition since the integral can easily be checked to be of order $\frac{1}{d}$ uniformly for $\varepsilon \in (0,1]$. From the definitions of α and τ, a straightforward but long calculation takes us to

$$\tau(\alpha)(\theta) = \frac{-4\varepsilon^{3d}d^2\sec^4\frac{\theta}{2}(\tan^{d-2}\frac{\theta}{2} - \tan^{-d-2}\frac{\theta}{2})}{(1+\varepsilon^{2d}(\tan^d\frac{\theta}{2} - \tan^{-d}\frac{\theta}{2})^2)^2}. \tag{3.5}$$

Plugging this into the expression

$$\|\mathcal{T}(w_\varepsilon)\|^2_{L^2(S^2)} = 2\pi\int_0^\pi |\tau|^2\sin\theta d\theta, \tag{3.6}$$

leads us to (2.2) of the proposition, with the integral being dominated now by values of θ of the order of ε. □

Remark 3.1. Related to w_ε is the map $u_\varepsilon : S^2 \to S^2$ given by

$$u_\varepsilon(z) = \varepsilon^d(z^d + \bar{z}^{-d}).$$

Calculations for u_ε show (cf. Proposition 2.1) that

$$-c_0 d\varepsilon^{2d} \leq E(u_\varepsilon) - 8\pi d \leq -C_0 d\varepsilon^{2d} \tag{3.7}$$

for universal $c_0, C_0 > 0$, and

$$c_1\varepsilon^{2d-1} \leq \|\mathcal{T}(u_\varepsilon)\|_{L^2(S^2)} \leq C_1\varepsilon^{2d-1} \tag{3.8}$$

for $c_1, C_1 > 0$ dependent only on d. This turns out to be useful in the theory of biharmonic maps, as may be described in [2].

---

[2] cf. the case $d = 1$ in, say, [1]

We now consider the significance of the maps w_ε to the harmonic map heat flow. As we see in the following proposition, it is possible to use these maps, with ε dependent on time, as barriers.

Proposition 3.2. *Let $\varepsilon : [0,\infty) \to (0,1]$ solve the ODE $\dot{\varepsilon} = -16d\varepsilon^{2d-1}$ with initial condition $\varepsilon(0) = 1$. Then*

$$\frac{\partial}{\partial t}\alpha_{\varepsilon(t)}(\theta) - \tau\left(\alpha_{\varepsilon(t)}\right)(\theta) \geq 0 \tag{3.9}$$

for $\theta \in [0, \frac{\pi}{2}]$, $t \geq 0$.

By symmetry, the left-hand side of (3.9) must be weakly negative for $\theta \in [\frac{\pi}{2}, \pi]$. The proof of the proposition is a straightforward calculation using the inequality $r^2(1 + \varepsilon^{2d}(r^d - r^{-d})^2) \geq \varepsilon^2/2$ valid for $r > 0$, $\varepsilon \in (0,1]$, $d \geq 1$, say. Note that neither this inequality nor the ODE of the proposition is sharp.

Remark 3.3. If we now have a smooth solution of (1.1) of the form

$$(\theta, \phi, t) \to (h(\theta, t), d\phi) \tag{3.10}$$

where $h : [0,\pi] \times [0,T] \to \mathbb{R}$ satisfies $h(0,\cdot) \equiv -\pi$, $h(\theta,\cdot) \equiv -h(\pi - \theta, \cdot)$ and $h(\theta, 0) \leq \alpha_1(\theta)$ for $\theta \in [0, \frac{\pi}{2}]$ then by virtue of Proposition 3.2, we may apply the comparison principle (see [1] for details in a very similar situation) to establish that

$$h(\theta, t) \leq \alpha_{\varepsilon(t)}(\theta)$$

for $\theta \in [0, \frac{\pi}{2}]$ and $t \in [0, T]$, with an analogous (reversed) inequality for $\theta \in [\frac{\pi}{2}, \pi]$. Similarly, if $h(\theta, 0) \in [-\pi, \pi]$ for $\theta \in [0, \pi]$, then the comparison principle tells us that $h(\theta, t) \in [-\pi, \pi]$ for $\theta \in [0, \pi]$ and $t \in [0, T]$.

Proof. (Theorem 2.3.) The work of Struwe [9] gives the existence of a weak solution u with $u(0) = u_0$, which is smooth on some maximal time interval $[0, T)$ with $T \in (0, \infty]$, and indeed smooth throughout space-time except possibly at finitely many points. There is no other smooth solution on $[0, T)$ with u_0 as initial map – see for example [9]. Since the initial map is of the form (3.1) with $\beta \equiv \alpha_1$, and $\alpha_1(\theta) = -\alpha_1(\pi - \theta)$, the uniqueness we have described guarantees that the flow will be of the form (3.10) for $t \in [0, T)$, with h satisfying $h(0, \cdot) \equiv -\pi$ and $h(\theta, \cdot) \equiv -h(\pi - \theta, \cdot)$ for $\theta \in [0, \frac{\pi}{2}]$. By the finiteness of the singular set, a flow of this form must be smooth away from the poles $\theta = 0$ and $\theta = \pi$ at time $t = T$ if $T < \infty$, and so by the bubbling description of singularities in the Struwe flow given in [9], the only way that u could fail to be smooth would be by bubbling at the poles. Meanwhile, since $h(\cdot, 0) \equiv \alpha_1$, Remark 3.3 tells us that $-\pi \leq h(\theta, t) \leq \alpha_{\varepsilon(t)}(\theta)$ for $\theta \in [0, \frac{\pi}{2}]$ and $t \in [0, T)$, where $\varepsilon(t) > 0$ is as in Proposition 3.2. In particular, this restricts the image of the flow near the pole $\theta = 0$ at finite time, ruling out the development of a bubble at finite time and forcing $T = \infty$. Note that our knowledge of the harmonic maps between 2-spheres, and the bubbling theory of Struwe then forces the flow limit u_∞ to be the constant map $(\theta, \phi) \to (0, d\phi)$.

Although this argument holds for all integral $d \geq 1$, we are assuming $d > 1$ in the theorem, and therefore the solution of the ODE for ε in Proposition 3.2 is

$$\varepsilon(t) = (1 + 32d(d-1)t)^{-\frac{1}{2(d-1)}}.$$

For $t \geq 1$, say, we have

$$\varepsilon(t) \geq \eta t^{-\frac{1}{2(d-1)}}, \tag{3.11}$$

for some $\eta = \eta(d) > 0$ which we may reduce during the proof. Meanwhile, for $0 \leq \theta \leq \varepsilon \leq 1$, we may estimate

$$\alpha_\varepsilon(\theta) \leq -\eta$$

for some small $\eta > 0$ independent of ε, and therefore, since $\alpha_{\varepsilon(t)}(\theta) \geq h(\theta, t)$ for this range of θ, and all $t \geq 0$, we have $|u(t) - u_\infty| \geq \eta > 0$ (for a possibly smaller η) on the disc defined by $\theta \leq \varepsilon(t)$. Consequently,

$$\|u(t) - u_\infty\|_{L^2(S^2)} \geq \eta\varepsilon(t),$$

which once combined with (3.11) is (2.3) of Theorem 2.3. In order to tackle (2.4) let us return to the definition of τ given in (3.3), and assume that $\beta(0) = -\pi$ and $\beta(\frac{\pi}{2}) = 0$. Multiplying this definition by $\sin^2\theta\,\beta'(\theta)$ and integrating from $\theta = 0$ to $\theta = \frac{\pi}{2}$, we obtain (after integrating by parts)

$$\int_0^{\frac{\pi}{2}} \sin^2\theta\,\tau\,\beta'(\theta)d\theta = \frac{1}{2}(\beta'(\frac{\pi}{2}))^2.$$

Recalling (3.2) and (3.6), Cauchy-Schwarz then tells us that

$$\left(\beta'(\frac{\pi}{2})\right)^2 \leq 2\int_0^{\frac{\pi}{2}} \sin\theta\,\tau\,\beta'(\theta)d\theta \leq CE^{\frac{1}{2}}\|\mathcal{T}\|_{L^2}.$$

When we apply this to flow u and its corresponding h, keeping in mind the relationship of $h(\cdot, t)$ and $\alpha_{\varepsilon(t)}$, we find that

$$\|\mathcal{T}(u(t))\|_{L^2(S^2)} \geq \eta\left(\frac{\partial h}{\partial\theta}(\frac{\pi}{2}, t)\right)^2 \geq \eta\left(\alpha'_{\varepsilon(t)}(\frac{\pi}{2})\right)^2 \geq \eta\varepsilon^{2d},$$

where the value of $\eta = \eta(d) > 0$ is allowed to decrease at each instance. This establishes (2.4). The final two parts of Theorem 2.3 follow easily from (2.4). Straightforward integration gives (2.5), whilst integrating the square of (2.4), keeping in mind (1.2), gives (2.6). □

Remark 3.4. Only in the case $d > 1$ does the solution $\varepsilon(t)$ of the ODE in Proposition 3.2 fail to decay exponentially as required for Theorem 2.3. However, $\alpha_{\varepsilon(t)}$ serves as a natural barrier even for $d = 1$, preventing finite time blow-up – cf. Chang-Ding [1].

References

[1] K.-C. Chang and W.-Y. Ding, *A result on the global existence for heat flows of harmonic maps from D^2 into S^2*. 'Nematics.' J.-M. Coron et al ed., Kluwer Academic Publishers (1990) 37–48.

[2] N. Course, Thesis due 2004, University of Warwick.

[3] W.-Y. Ding and G. Tian, *Energy identity for a class of approximate harmonic maps from surfaces.* Comm. Anal. Geom. **3** (1995) 543–554.

[4] J. Eells and L. Lemaire, *A report on harmonic maps.* Bull. London Math. Soc. **10** (1978) 1–68.

[5] R. Gulliver and B. White, *The Rate of Convergence of a Harmonic Map at a Singular Point.* Math. Ann. **283** (1989) 539–549.

[6] J. Qing, *On singularities of the heat flow for harmonic maps from surfaces into spheres.* Comm. Anal. Geom. **3** (1995) 297–315.

[7] L. Simon, *Asymptotics for a class of nonlinear evolution equations, with applications to geometric problems.* Ann. Math. **118** (1983) 525–571.

[8] L. Simon, *Theorems on the Regularity and Singularity of Minimal Surfaces and Harmonic Maps.* 'Geometry and Global Analysis.' T. Kotake et al ed., MSJ report, Sendai, Japan (1993) 111–145.

[9] M. Struwe, *On the evolution of harmonic mappings of Riemannian surfaces.* Comment. Math. Helv. **60** (1985) 558–581.

[10] P.M. Topping, *Rigidity in the Harmonic Map Heat Flow.* J. Differential Geometry **45** (1997) 593–610.

[11] P.M. Topping, *Repulsion and quantization in almost-harmonic maps, and asymptotics of the harmonic map flow.* To appear, Annals of Math.

[12] C.-Y. Wang, *Bubble phenomena of certain Palais-Smale sequences from surfaces to general targets.* Houston J. Math. **22** (1996) 559–590.

Peter Topping
Mathematics Institute
University of Warwick
Coventry, CV4 7AL, UK
e-mail: `topping@maths.warwick.ac.uk`
URL: `http://www.maths.warwick.ac.uk/~topping`

Part II

Evolution of Maps and Metrics

Progress in Nonlinear Differential Equations
and Their Applications, Vol. 59, 45–65

Heat Flow into Spheres for a Class of Energies

Norbert Hungerbühler

Abstract. Let M and N be compact smooth Riemannian manifolds without boundaries. Then, for a map $u : M \to N$, we consider a class of energies which includes the popular Dirichlet energy and the more general p-energy. Geometric or physical questions motivate to investigate the critical points of such an energy or the corresponding heat flow problem. In the case of the Dirichlet energy, the heat flow problem has been intensively studied and is well understood by now. However, it has turned out that the case of the p-energy ($p \neq 2$) is much more difficult in many respects. We give a survey of the known results for the p-harmonic flow and indicate how these results can be extended to a larger class of energy types by using Young measure techniques which have recently been developed for quasilinear problems.

1. Introduction

1.1. The p-energy

Let M and N be smooth Riemannian manifolds without boundary of dimension m and n respectively. M and N are equipped with Riemannian metrics γ and g. The p-energy of a differentiable map $u : M \to N$ is defined to be

$$E_p(u) := \frac{1}{p}\int_M |Du|^p d\mu.$$

Here, $Du : T_xM \to T_{u(x)}N$ is the differential of u to which we associate the p-energy density

$$e_p(u) := \frac{1}{p}|Du|^p = \frac{1}{p}(\text{trace}((Du)^* Du))^{p/2},$$

where '$*$' denotes the adjoint with respect to the underlying inner product on the respective tangent spaces T_xM and $T_{u(x)}N$. And, of course, $d\mu$ is the measure on M associated with the metric on M. In local coordinates, the p-energy can be expressed by the following formula:

$$E_p(u) := \frac{1}{p}\int_M (\gamma^{\alpha\beta}(g_{ij} \circ u)\partial_\alpha u^i \partial_\beta u^j)^{p/2}\sqrt{\gamma}\, dx\,.$$

Here, $\sqrt{\gamma} = \sqrt{|\det(\gamma_{\alpha\beta})|}$, and we adopt the usual summation convention, i.e., we automatically sum over Latin indices from 1 to n and over Greek indices from 1 to m. By Nash's embedding theorem, we always may assume that N is isometrically embedded in some Euclidean space $\mathbb{R}^k$. In that case, the p-energy density is given by

$$\frac{1}{p}(\gamma^{\alpha\beta}\partial_\alpha u \cdot \partial_\beta u)^{p/2}$$

where $u : M \to N \subset \mathbb{R}^k$ is viewed as a map into $\mathbb{R}^k$ and '$\cdot$' is the Euclidean inner product in $\mathbb{R}^k$.

We remark, that for $p = 2$ the p-energy coincides with the well-known Dirichlet energy. Moreover, as it is easy to see, the p-energy is conformally invariant iff $m = p$. This fact allows to switch between conformal charts.

1.2. The first variation of the p-energy

Given a class F of mappings from M to N defined by boundary conditions on ∂M (if there is a boundary) and possibly additional topological conditions, e.g., a homotopy class, one may try to minimize the p-energy within the class F. One may hope that the solution of this minimization problem is a map u in F with particular analytic and geometric properties. In physics, the p-energy for maps into the sphere or into a Grassmannian manifold may serve as a model for liquid crystals. Stationary points of the energy then correspond to physical equilibria of the liquid.

Necessarily, in a minimum u, the first variation of the p-energy must vanish at u for all variations of u with compact support in the interior of M. Let us consider variations $u_t : M \to N$ of $u = u_0$ for $|t|$ small such that $u_t(x)$ is of class C^1 in the variables (x, t) and $u_t(x) = u(x)$ for all t and all x outside some compact subset $K \subset M$ which is contained in the domain U of a coordinate system $x^1, \ldots, x^m$ on M with range Ω and is mapped into the domain of a fixed coordinate system $y^1, \ldots, y^m$ on N by all u_t. Now, the support of the initial vector field of the variation

$$V(x) := \frac{d}{dt}u_t(x)\bigg|_{t=0} \in T_{u(x)}N$$

is contained in K. We also assume that V is of the class C^1. It is no restriction to assume $u_t^i(x) = u^i(x) + tV^i(x)$. If u is C^2, the usual computation yields that vanishing first variation is equivalent to

$$\frac{1}{\sqrt{\gamma}}\frac{\partial}{\partial x^\beta}\left(\sqrt{\gamma}\left(\gamma^{\alpha\beta}g_{ij}\frac{\partial u^i}{\partial x^\alpha}\frac{\partial u^j}{\partial x^\beta}\right)^{\frac{p}{2}-1}\gamma^{\alpha\beta}\frac{\partial u^l}{\partial x^\alpha}\right) = \tag{1}$$

$$= -\left(\gamma^{\alpha\beta}g_{ij}\frac{\partial u^i}{\partial x^\alpha}\frac{\partial u^j}{\partial x^\beta}\right)^{\frac{p}{2}-1}\gamma^{\alpha\beta}\Gamma^l_{ij}\frac{\partial u^i}{\partial x^\alpha}\frac{\partial u^j}{\partial x^\beta}$$

in U for $l = 1, \ldots, n$. Here, we used the Christoffel symbols

$$\Gamma^l_{ij} = \frac{1}{2}g^{lk}(g_{ik,j} - g_{ij,k} + g_{jk,i})$$

of the metric g with respect to the coordinates chosen on N. Thus, the right-hand side can be interpreted as

$$A(u)(Du, Du)(pe_p(u))^{1-2/p}$$

where $A(u)(\cdot,\cdot)$ denotes the second fundamental form of N. The operator on the left-hand side of (1) is called p-Laplace operator related to the manifolds M and N and is denoted by Δ_p, i.e., Δ_2 is simply the Laplace-Beltrami operator of the manifold M and does not depend on N. Δ_2 is a linear elliptic diagonal operator in divergence form. For $p > 2$ ($0 < p < 2$) the operator is degenerate (singular) at points $Du = 0$.

The right-hand side of (1) is for $p = 2$ a quadratic form in the first derivatives of u with coefficients depending on u. These strong nonlinearities are caused by the non-Euclidean structure of the target manifold N and cannot be removed by special choices of coordinates on N unless N is locally isometric to Euclidean space $\mathbb{R}^n$. But even in this case the space of mappings from M to N does not possess a natural linear structure unless N itself is a linear space. In general, the right-hand side of (1) is of the order of the p-th power of the gradient of u.

A C^2-solution $u : M \to N$ of (1) is called p-harmonic map.

Using variations in the *domain*, i.e., variations of the form

$$u_t(x) = u(x + t\zeta(x))$$

where $\zeta = (\zeta^1, \dots, \zeta^m)$ with each $\zeta^j \in C_0^\infty(B_\rho(y))$, we obtain from $\frac{d}{dt}u_t(x)\big|_{t=0} = 0$ an equation whose classical solutions contain the set of the p-harmonic maps. However its weak solutions, which are called "stationary p-harmonic maps", need not contain the set of weak solutions of (1). In particular, p-energy minimizing maps are both, weakly and stationary p-harmonic. Weakly p-harmonic maps admit in general far worse singularities than the energy minimizing maps. Many results are known about stationary p-harmonic maps. See, e.g., Duzaar and Fuchs [9], Fuchs [18] and [19], as well as Strzelecki [41] and [40].

If we think of N as being isometrically embedded in some $\mathbb{R}^k$, we consider the following variations: Let $S \subset \mathbb{R}^k$ be a tubular neighborhood of N and $\pi_N : S \to N$ the (smooth) nearest-neighbor projection. Denote $T_pN \subset T_p\mathbb{R}^k$ the tangent space to N at a point $p \in N$. Let $\varphi \in C_0^1(M, \mathbb{R}^k)$ satisfy

$$\varphi(x) \in T_{u(x)}N$$

for all $x \in M$ and φ having compact support in a single coordinate chart of M. Then φ induces a C^1-variation at $u : M \to N \subset \mathbb{R}^k$:

$$u_t = \pi_N \circ (u + t\varphi)\,.$$

Now, the first variation of the p-energy is given by

$$\begin{aligned} \frac{d}{dt}E_p(u_t)\bigg|_{t=0} &= \int_M \left(\gamma^{\alpha\beta}\frac{\partial u}{\partial x^\alpha}\cdot\frac{\partial u}{\partial x^\beta}\right)^{\frac{p}{2}-1}\gamma^{\alpha\beta}\frac{\partial u}{\partial x^\alpha}\cdot\frac{\partial\varphi}{\partial x^\beta}\sqrt{\gamma}\,dx = \\ &= -\int_M \Delta_p u\cdot\varphi\sqrt{\gamma}\,dx \end{aligned}$$

where Δ_p is the p-Laplace operator related to M and $\mathbb{R}^k$, i.e.,

$$\Delta_p u = \frac{1}{\sqrt{\gamma}} \frac{\partial}{\partial x^\beta} \left(\sqrt{\gamma} \left(\gamma^{\alpha\beta} \frac{\partial u^j}{\partial x^\alpha} \frac{\partial u^j}{\partial x^\beta} \right)^{\frac{p}{2}-1} \gamma^{\alpha\beta} \frac{\partial u}{\partial x^\alpha} \right).$$

Thus, the first variation vanishes at a C^2 map u if and only if

$$\Delta_p u \perp T_u N . \tag{2}$$

We can make (2) more explicit by introducing a local orthonormal frame $\nu_{n+1}, \ldots, \nu_k$ for $(T_p N)^\perp$, the orthogonal complement of $T_p N$ in $\mathbb{R}^k$. Then, by (2) there exist scalar functions $\lambda^{n+1}, \ldots, \lambda^k$ such that

$$-\Delta_p u = \sum_{l=n+1}^{k} \lambda^l (\nu_l \circ u) . \tag{3}$$

Multiplying (3) by $\nu_i \circ u$ (i fixed) and using the fact that $\frac{\partial u}{\partial x^\alpha} \cdot \nu_l(u) = 0$ for all α, we obtain

$$\lambda^i = \left(\gamma^{\alpha\beta} \frac{\partial u^j}{\partial x^\alpha} \frac{\partial u^j}{\partial x^\beta} \right)^{\frac{p}{2}-1} \gamma^{\alpha\beta} \frac{\partial u^j}{\partial x^\alpha} \frac{\partial \nu_i^j(u)}{\partial x^\beta} . \tag{4}$$

The Euler-Lagrange equations of the p-energy can also be formulated in the following coordinate free form:

$$e(u)\tau(u) + (1 - \frac{2}{p}) Du \operatorname{grad}_\gamma e(u) = 0 . \tag{5}$$

Here, $\tau(u)$ denotes the tension field of u: $\tau(u) = \operatorname{trace}_\gamma \nabla Du$, and ∇ denotes the pull-back covariant derivative in the bundle $T^*M \otimes u^{-1}TN$. For $p = 2$ the equation reduces simply to

$$\tau(u) = 0.$$

As an application, we see, that the identity mapping $\mathrm{id}_M : M \to M$ of a Riemannian manifold is p-harmonic: Since $D(\mathrm{id}_M)$ has constant coefficients with respect to all coordinate systems on M, $\nabla D(\mathrm{id}_M) = 0$ and the first term in (5) vanishes. On the other-hand, the p-energy density $e(\mathrm{id}_M)$ is constant and hence the gradient in the second term disappears, too. We remark that in general id_M is not p-energy minimizing within its homotopy class (see Eells-Lemaire [11]).

The natural space to work in when considering questions concerning the p-energy is the nonlinear Sobolev space

$$W^{1,p}(M,N) := \{ f \in W^{1,p}(M, \mathbb{R}^k);\ f(x) \in N \text{ for } \mu\text{-almost all } x \in M \}.$$

Notice, that if both M and N are compact, then different embeddings of N in $\mathbb{R}^k$ give rise to homeomorphic spaces $W^{1,p}(M,N)$. For an intrinsic definition of the space $W^{1,p}(M,N)$ see Federer [17]. If N is isometrically embedded in $\mathbb{R}^k$, then the formulas defining the p-energy for C^1 functions make sense also for functions in $W^{1,p}(M,N)$. If a function $u \in W^{1,p}(M,N)$ is a weak solution of the Euler-Lagrange equations we discussed previously, then it is called a weakly p-harmonic map.

The space $H^{1,p}(M,N)$ defined as the closure of the class of smooth functions from M to N in the $W^{1,p}$-norm is contained in $W^{1,p}(M,N)$ but does not coincide with the latter space in general (this fact gives rise to the so called "gap phenomenon" of Hardt-Lin [21]). This important observation was first made by Schoen and Uhlenbeck: see Eells and Lemaire [11] as a main reference. However, we have $H^{1,p}(M,N) = W^{1,p}(M,N)$ if $\dim(M) = p$ (see Schoen and Uhlenbeck [39], Bethuel [2] or Bethuel and Zheng [3]). The gap phenomenon limits the possibilities to approximate the p-harmonic flow, which we will discuss next, by solutions with smooth initial data.

1.3. The p-harmonic flow

A basic existence problem for p-harmonic maps is the homotopy problem:

Given a map $u_0 : M \to N$ is there a p-harmonic map u homotopic to u_0?

This question was first answered in the case $p = 2$ of harmonic maps. The answer is affirmative if the sectional curvature K^N of N is non-positive (see Eells-Sampson [12]), or – in case of a two-dimensional surface M – if the second fundamental group of N is trivial: $\pi_2(N) = 0$ (see Lemaire [30] and Sacks-Uhlenbeck [38]). Eells and Wood destroyed the hope for a more general theorem by the following counterexample in [13] by showing that, if $u : T^2 \to S^2$ is 2-harmonic, then $\deg u \neq \pm 1$. Another counterexample has been given by Lemaire in [30]. Thus, in general the attempt to solve the homotopy problem by minimizing E within a given homotopy class will fail: Homotopy classes are not weakly closed in $W^{1,p}(M,N)$ generally. Thus, it may be difficult or impossible to solve the homotopy problem for p-harmonic maps by direct variational methods. As the key idea to get around this difficulty, Eells and Sampson proposed in [12] to study the heat flow related to the 2-energy:

$$u_t - \Delta_2 u = A(u)(\nabla u, \nabla u)_M \qquad \text{on } M \times [0,\infty[\tag{6}$$

with initial and boundary data

$$u = u_0 \text{ at } t = 0 \text{ and on } \partial M \times [0,\infty[$$

for maps $u : M \times [0,\infty[\to N \subset \mathbb{R}^k$. Here, $A(u) : T_uN \times T_uN \to (T_uN)^\perp$ is the second fundamental form of N. The idea behind this strategy is of course that a continuous deformation $u(\cdot,t)$ of u_0 will remain within the given homotopy class. Since (6) may be interpreted as the L^2-gradient flow for the 2-energy, one may hope that the solution $u(\cdot,t)$ for $t \to \infty$ will come to a rest at some critical point of E_2 that is a harmonic map. For target manifold N, satisfying the geometric restrictions mentioned above, this program has been applied with success for $p = 2$.

Another approach to the homotopy problem for p-harmonic maps has been given by Duzaar and Fuchs in [10]: they extended the Eells-Sampson result to the case $p \in [2,\infty)$ by using an asymptotic analysis of the not degenerate energy $\int_M (\varepsilon + |Du|^2)^{p/2} d\mu$, $\varepsilon > 0$.

Corresponding to the harmonic flow, the p-harmonic flow is described by

$$u_t - \Delta_p u = (pe(u))^{1-\frac{2}{p}} A(u)(Du, Du) \qquad \text{on } M \times [0,\infty[\tag{7}$$

with initial data

$$u = u_0 \text{ at } t = 0$$

for maps $u : M \times [0, \infty[\to N \subset \mathbb{R}^k$.

We now want to briefly describe the known results concerning the p-harmonic flow. We restrict this presentation to the case $p \neq 2$.

The p-harmonic flow was first considered in [6] and [22]. There, the following result has been proved for the p-harmonic flow into spheres, i.e., for

$$\partial_t u - \Delta_p u = p e_p(u) u \qquad \text{on } M \times [0, \infty[\tag{8}$$

$$u(\,\cdot\,, 0) = u_0 \qquad \text{on } M \tag{9}$$

$$|u| = 1 \qquad \mu\text{-a.e. on } M \times [0, \infty[. \tag{10}$$

Theorem 1. *For initial data $u_0 \in W^{1,p}(M, S^n)$, $p \geq 2$, there exists a global weak solution u to the equation (8)–(10). This solution is weakly continuous in $t > 0$ with values in $W^{1,p}(M)$, i.e., for any test function $g \in C^\infty(M)$, $h_1(t) = \int_M u \cdot g \, d\mu$ and $h_2(t) = \int_M Du \cdot Dg \, d\mu$ are in $C^{0,\frac{1}{2}}(\mathbb{R}_+)$. Furthermore for almost every $t \geq 0$ this solution satisfies the energy inequality*

$$\int_0^t \|\partial_t u\|^2_{L^2(M)} dt + E_p(u(t)) \leq E_p(u_0) \,. \tag{11}$$

The technique was the same as in the corresponding case of the harmonic flow which was solved by Chen in [5]: The idea is a penalization technique to approximate the p-harmonic flow, which will be explained and used in Section 2 for a more general class of energies. The monotonicity of the p-Laplace operator allows to solve the approximating equations by Minty's trick alone. Nonetheless, to pass to the limit in the penalized equations some stronger compactness results on the p-Laplacien are needed. In the original work, this is based on the uniform monotonicity of the p-Laplace operator:

Theorem 2. *For $k = 1, 2, \ldots$, let $f_k : M \times [0, T] \to \mathbb{R}^l$ satisfy the equation*

$$\partial_t f_k - \Delta_p f_k = g_k, \quad \text{on } M \times [0, T]$$

in the sense of distributions. Assume that

1. *$\{f_k\}_{k \in \mathbb{N}}$ is bounded in $L^\infty(0, T; W^{1,p}(M, \mathbb{R}^l))$,*
2. *$\{\partial_t f_k\}_{k \in \mathbb{N}}$ is bounded in $L^2(0, T; L^2(M, \mathbb{R}^l))$, and that*
3. *$\{g_k\}_{k \in \mathbb{N}}$ is bounded in $L^1(0, T; L^1(M, \mathbb{R}^l))$.*

Then, $\{f_k\}_{k \in \mathbb{N}}$ is precompact in $L^q(0, T; W^{1,q}(M, \mathbb{R}^l))$ for each $1 \leq q < p$.

In Section 2 we offer a new approach to this compactness problem which involves some Young measure techniques, and which does not use strict or even uniform monotonicity properties of the operator. This allows in particular to obtain results for the flow of energies which are merely convex.

Theorem 1 was later proved for $1 < p < 2$ by Liu in [32], [33] and by Misawa in [37]. The p-harmonic flow from a unit ball in $\mathbb{R}^m$ into $S^1 \subset \mathbb{R}^2$ was also considered by Courilleau and Demengel in [7]. There, also non-uniqueness of the p-harmonic flow with weakly p-harmonic, but not stationary harmonic initial data was proven, a result, however, which is already contained in [44] and even before that in [25].

The p-harmonic flow equation has two nonlinear terms: The p-Laplace term, and the right-hand side involving the second fundamental form. The previously described compactness techniques allow to pass to the limit in the p-Laplace term only. The big problem, which is unsolved to the present day, is to pass to the limit in the nonlinear term on the right-hand side of the equation. The sphere as target manifold has enough symmetries to rewrite the equations in such a way, that the right-hand side simply vanishes (compare Section 2). A similar trick can be played to prove existence of the p-harmonic flow if the target manifold N is a homogeneous space with a left invariant metric.

Theorem 3 ([23]). *For $2 < p$ there exists a global weak solution of the p-harmonic flow between Riemannian manifolds M and N for arbitrary initial data having finite p-energy in the case when the target N is a homogeneous space with a left invariant metric. The solution $u : M \times [0, \infty[\to N$ satisfies the energy inequality*

$$\frac{1}{2}\int_0^T \int_M |\partial_t u|^2 dt\, d\mu + \frac{1}{p}\int_M |Du(T)|^p d\mu \le \frac{1}{p}\int_M |Du(0)|^p d\mu \tag{12}$$

for almost all $T > 0$.

In this case of a homogeneous space as target, the equation of the p-harmonic flow can be reformulated such that the right-hand side has a divergence structure (see [24] and [23]). In [23], a technique different from the penalization technique was used to approximate the equation, namely a time discretization. The reason is, that the penalization does not preserve the special structure of N.

The analogous result for $1 < p < 2$ was proven in [31]. There, the case of a modified p-energy is treated, namely that of the p-energy 'with potential', i.e., $E(u) = E_p - \int_M H(u) d\mu$ for a Lipshitz function H.

Another particular case which has been solved is the conformal case $p = m = \dim M$:

Theorem 4 ([26]). *For given initial value $u_0 \in W^{1,m}(M, N)$ there exists a weak solution $u : M \times [0, \infty[\to N$ of the m-harmonic flow*

$$\begin{aligned} u_t - \Delta_m u \quad &\perp \quad T_u N \\ u(\cdot\,, 0) \quad &= \quad u_0\,. \end{aligned}$$

u satisfies the energy inequality (12) and is in $W^{1,m}(M)$ weakly continuous in time. There exists a set $\Sigma = \cup_{k=1}^K \Sigma_k \times \{T_k\}$, $\Sigma_k \subset M$, $0 < T_k \le \infty$, such that on every open set $\Omega \subset M \times [0, \infty[$ with $\operatorname{dist}(\Omega, (M \times \{0\}) \cup \Sigma) = \mu > 0$ there holds $\|\nabla u\|_{C^{0,\beta}(\Omega)} \le C$ for some constants C (depending on m, $E_m(u_0)$, M, N and μ) and $\beta \in]0, 1[$ (depending on m, M and N). There exists $\varepsilon_1 > 0$ such that the number K of singular times is a priori bounded in terms of the initial energy, $K \le \varepsilon_1^{-1} E_m(u_0)$, and the singular points (x, T_k) are characterized by the condition

$\limsup_{t\nearrow T_k} E_m(u(t), B_R(x)) \geq \varepsilon_1$ *for any* $R > 0$. *At every singular time* T_k *the decrease of the* m*-energy is at least* ε_1*:*

$$E_m(u(T_k)) \leq \liminf_{t\nearrow T_k} E_m(u(t)) - \varepsilon_1.$$

In order to get rid of the nonlinear side condition, that the solution has to take values in the sphere, Hamilton's geometric technique from [20] was used, namely that of a total geodesic embedding of the target N in some $\mathbb{R}^k$. I.e., the Euclidean metric in $\mathbb{R}^k$ is deformed to a new metric h such that

1. N is still embedded isometrically,
2. the metric h equals the Euclidean metric outside a large ball,
3. there exists an involutive isometry $\iota : T \to T$ on a tubular neighborhood T of N corresponding to multiplication by -1 in the orthonormal fibers of N and having precisely N for its fixed point set.

A h-geodesic curve γ connecting x, $y \in N$ (x, y close enough) will always be contained in N. This follows from the (local) uniqueness of geodesics and the fact that with γ the curve $\iota \circ \gamma$ is another geodesic joining x and y.

The idea is that, if the initial data are smooth, then a short time solution u of the p-harmonic flow is unique which forces it to stay in N (since otherwise, u and $\iota \circ u$ were two different solutions).

The crucial point in the conformal case is an energy concentration lemma:

Theorem 5 ([26]). *If* $\dim(M) = p$ *and* $u \in C^2(M \times [0, T[; N)$ *is a solution of the* m*-harmonic flow then there exist constants* c, $\varepsilon_0 > 0$ *which only depend on the geometry of the manifolds* M *and* N, *and there exists a time* $T_0 \in]0, T[$ *which depends in addition on* u_0, *with the following properties: If the initial local energy satisfies*

$$\sup_{x\in M} E_m(u(0), B_{2R}(x)) < \varepsilon_0$$

then it follows

$$E_m(u(t), B_R(x)) \leq E_m(u(0), B_{2R}(x)) + c\, E_0^{1-\frac{1}{p}} \frac{t}{R^m} \tag{13}$$

for all $(x, t) \in M \times [0, T_0]$. *Here* E_0 *denotes the initial energy, and* $E_m(u(t), B)$ *is the* m*-energy of* $u(t)$ *in a region* $B \subset M$.

This lemma allows to control energy concentration of the flow for short time which leads to short time existence and (local) smoothness. At times, where the energy concentrates, the flow develops a singularity and can be restarted. It is not known whether for $m \neq 2$ the m-harmonic flow develops singularities in finite time, in contrast to the harmonic flow, where this is confirmed by the example of Chang, Ding and Ye in[4].

Technically, the proof of Theorem 4 is carried out by two approximation steps: In a first step, only smooth initial data are considered. At the same time, the p-energy is regularized by $E_{p,\varepsilon} = \int_M (\varepsilon + |Du|^2)^{p/2} d\mu$. In a second step, arbitrary initial data are approximated by smooth functions (observe, that by Bethuel-Zheng [3] $C^\infty(M,N)$ is dense in $W^{1,p}(M,N)$). Misawa improved Theorem 4 in [34] by showing, that, as it is the case for the harmonic flow, the energy concentration set Σ is discrete not only in time, but also in space.

The p-harmonic flow in the case of non-positive sectional curvature of the target manifold was treated by Fardoun and Regbaoui in [14] and [16]:

Theorem 6. *If* $\mathrm{Riem}_N \leq 0$ *and* $u_0 \in C^{2,\alpha}(M,N)$, *then for* $p > 1$:

1. *There exists a unique global weak solution* u *of the* p*-harmonic flow from* M *to* N *such that* $\partial_t u \in L^2([0,\infty[\times M)$ *and* $u, Du \in C^\beta([0,\infty[\times M)$ *for some* $\beta \in]0,1[$. *Moreover, the solution satisfies the energy inequality* (12) *for all* $t > 0$.
2. *There exists a sequence* $t_k \to \infty$ *such that* $u(t_k,\cdot)$ *converges in* $C^{1,\beta'}(M,N)$, *for all* $\beta' < \beta$, *to a weakly* p*-harmonic map* $u_\infty \in C^{1,\beta}(M,N)$.

In particular, this results settles the homotopy question in this case. The line of proof goes along regularization of the p-energy, and a Bochner formula for the energy density (see [14] for details).

The same result for $p > 2$ was shown earlier by Misawa in [35] under the assumption that the image of u_0 is contained in a geodesic ball, and in [36] without that latter assumption.

The p-harmonic flow for the case of small initial data has been treated by Fardoun and Regbaoui in [15]:

Theorem 7. *For any* $p > 1$ *and* $K > 0$ *there exists an* $\varepsilon_0 > 0$ *depending on* K, M, N *and* p *such that if* $u_0 \in C^{2,\alpha}(M,N)$, $0 < \alpha < 1$, *with* $E_p(u_0) \leq \varepsilon_0$ *and* $\|Du_0\|_{L^\infty}(M) \leq K$, *the following is true:*

1. *There exists a unique global weak solution* u *of the* p*-harmonic flow from* M *to* N *such that* $\partial_t u \in L^2([0,\infty[\times M)$ *and* $u, Du \in C^\beta([0,\infty[\times M)$ *for some* $\beta \in]0,1[$. *Moreover, the solution satisfies the energy inequality* (12) *for all* $t > 0$.
2. *There exists a sequence* $t_k \to \infty$ *such that* $u(t_k,\cdot)$ *converges in* $C^{1,\beta'}(M,N)$, *for all* $\beta' < \beta$, *to a weakly* p*-harmonic map* $u_\infty \in C^{1,\beta}(M,N)$. *Moreover there exists* $\bar{\varepsilon}_0 > 0$ *depending on* K, M, N *and* p *such that if in addition* $E_p(u_0) \leq \bar{\varepsilon}_0$, *then* u_∞ *is a constant map.*

This result of Fardoun and Regbaoui concludes our survey on the p-harmonic flow.

2. Heat flow into spheres for a class of energies

In this section, our target manifold N is the unit sphere $S^n \subset \mathbb{R}^{n+1}$. In order to keep the formulas short, we restrict this presentation to the case $M = T = \mathbb{R}^m/\mathbb{Z}^m$ of a flat torus as domain manifold. The modifications necessary for a general compact smooth Riemannian manifold without boundary are purely technical and straightforward.

The class of energy functionals we want to investigate here is of the form

$$E(u) = \int_T \rho(|Du|)dx$$

for functions $u : T \to S^n \subset \mathbb{R}^{n+1}$. Here $\rho : \mathbb{R}_{\geq 0} \to \mathbb{R}$ is supposed to be continuously differentiable, convex, and satisfies the coercivity and growth conditions

$$C_1\xi^p \leq \rho(\xi) \qquad \text{for all } \xi \geq 0 \tag{14}$$

$$0 \leq \rho'(\xi) \leq C_2\xi^{p-1} \qquad \text{for all } \xi \geq 0 \tag{15}$$

for two positive constants C_1 and C_2, and for a given $p \geq 2$. The energy flow related to this energy is described by the following equation

$$\partial_t u - \operatorname{div}\Big(\rho'(|Du|)\frac{Du}{|Du|}\Big) \perp T_u S.$$

We use the shorthand notation

$$\sigma(Du) := \rho'(|Du|)\frac{Du}{|Du|}$$

and hence the previous equation can be rewritten in the form

$$\partial_t u - \operatorname{div}\sigma(Du) = \lambda u$$

for a function $\lambda(x,t)$. Observe, that by multiplying this equation by u, we find

$$\lambda = \sigma(Du) : Du.$$

Here, $A : B := \operatorname{trace}(AB^{\mathrm{T}})$ denotes the usual inner product of two matrices of the same type. So, the final form of the heat equation we want to solve, is

$$\partial_t u - \operatorname{div}\sigma(Du) = u\,\sigma(Du) : Du \qquad \text{on } T \times \mathbb{R}_{>0} \tag{16}$$

$$u(\cdot, 0) = u_0 \qquad \text{on } T. \tag{17}$$

We suppose, that the initial data u_0 have finite energy, i.e., $u_0 \in W^{1,p}(T, S^n)$.

2.1. Approximation by penalization

We construct a family of approximating problems by the classical penalization method. To this end, we consider the energies

$$E_k(u) := E(u) + k\int_T \chi(|u|)dx$$

for functions $u : T \to \mathbb{R}^{n+1}$, and for $k \in \mathbb{N}$. The smooth function $\chi : \mathbb{R}_{\geq 0} \to \mathbb{R}$ is chosen in such a way, that $0 \leq \chi \leq 1$, $\chi = 1$ outside the interval $[\frac{1}{2}, \frac{3}{2}]$, $\chi(x) = 0$ if and only if $x = 1$, and that χ' changes sign only once (see Figure 1). The idea is, that we abandon the restriction that u takes values in the target S^n, but that (in terms of energy) with increasing k, it becomes more and more favorable for the function u to take values close to the sphere S^n.

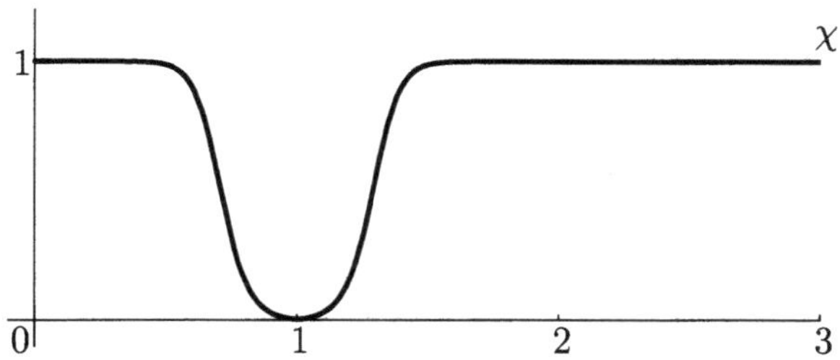

FIGURE 1. The function χ

The heat flow equation for the penalized energy is

$$\partial_t u - \operatorname{div} \sigma(Du) + k\chi'(|u|)\frac{u}{|u|} = 0. \tag{18}$$

The goal is now to prove existence of a global weak solution for this penalized energy flow with the given initial data (17).

Lemma 8. *Let $k \in \mathbb{N}$ and $u_0 \in W^{1,p}(T, S^n)$. Then, (17)–(18) possesses a weak solution $u_k \in L^\infty(0,\infty; W^{1,p}(T, \mathbb{R}^{n+1}))$ which is subject to the energy inequality*

$$\int_0^\tau \|\partial_t u_k\|^2_{L^2(T)} dt + E_k(u_k(\tau)) \le E(u_0) \tag{19}$$

for almost all $\tau \ge 0$.

Proof. $W^{1,p}(T, \mathbb{R}^{n+1})$ is separable and possesses therefore a Galerkin base $\{w_i\}_{i\in\mathbb{N}}$, $w_i \in W^{1,p}(T, \mathbb{R}^{n+1})$ in the sense, that for every $w \in W^{1,p}(T, \mathbb{R}^{n+1})$ there exist coefficients $c_i^{(j)}$ such that

$$w^{(j)} := \sum_{i=1}^{j} c_i^{(j)} w_i \to w \quad \text{in } W^{1,p}(T) \text{ as } j \to \infty.$$

It is convenient to choose the w_i smooth and L^2-orthonormal. For fixed $j \in \mathbb{N}$, we make the ansatz

$$u^{(j)}(t) := \sum_{i=1}^{j} c_i^{(j)}(t) w_i$$

for a solution of the approximating system

$$\int_T \Big(\partial_t u^{(j)} w_l + \sigma(Du^{(j)}) : Dw_l + k\chi'(|u^{(j)}|)\frac{u^{(j)}}{|u^{(j)}|}\, w_l\Big) dx = 0 \quad \text{for } l = 1, \dots, j \tag{20}$$

with initial values

$$u^{(j)}(0) = u_0^{(j)} \to u_0 \quad \text{in } W^{1,p}(T) \text{ for } j \to \infty. \tag{21}$$

This is a system of j ordinary differential equations for the coefficients $c_l^{(j)}(t)$, $l = 1, \dots, j$. It is easy to verify that this system satisfies the hypotheses of the existence theorem of Picard-Lindelöf and that therefore a solution exists on a time interval $]0,\tau(j)[$. Since, by construction, $u^{(j)}$ is an element of $\operatorname{span}(w_1,\dots,w_j)$, equation (20) holds with $u^{(j)}$ in place of w_l, and hence, after integration over a time interval $]0,\tau[$, $\tau\le\tau(j)$, one gets

$$\frac{1}{2}\|u^{(j)}(\tau)\|^2_{L^2(T)}+\int_0^\tau\int_T\sigma(Du^{(j)}):Du^{(j)}dxdt$$
$$+\int_0^\tau\int_T k\chi'(|u^{(j)}|)\frac{u^{(j)}}{|u^{(j)}|}u^{(j)}dxdt=\frac{1}{2}\|u^{(j)}(0)\|^2_{L^2(T)} \tag{22}$$

The first term in (22) is equal to $\frac{1}{2}|c^{(j)}(\tau)|^2_{\mathbb{R}^j}$. By (15), the second term is non-negative. Since $|\chi'(t)|t\le\frac{3}{2}\max|\chi'|$ for $t\ge 0$, the third term, in absolute value, is bounded by $C\tau$ for a constant C. This shows, that the functions $c_j^{(j)}(t)$ are bounded on $[0,\tau(j)[$, and hence, by (20), the same is true for $\frac{d}{dt}c_j^{(j)}(t)$. Therefore, the existence interval is both open and closed and hence we obtain a global solution $u^{(j)}$ of (20).

Similarly, with $\partial_t u^{(j)}$ in place of w_l in (20), we obtain

$$\int_0^\tau\|\partial_t u^{(j)}\|^2_{L^2(T)}dt+E_k(u^{(j)}(\tau))=E_k(u_0^{(j)})\le C\quad\text{for all } j \text{ and } \tau\ge 0. \tag{23}$$

Observe, that

$$E_k(u)\ge E(u)=\int_T\rho(|Du|)dx\ge C_1\int_T|Du|^p dx$$

by the coercivity assumption (14) on ρ. Therefore, by (23), the sequence $\{u^{(j)}\}_j$ is bounded in $L^\infty(0,\infty;W^{1,p}(T))$ On the other hand, still by (23), the sequence $\{\partial_t u^{(j)}\}_j$ is bounded in $L^2(T\times]0,\infty[)$. Aubin's lemma therefore implies that there exists a subsequence $\{u^{(j')}\}_{j'}$ and a function u such that

$$u^{(j')}\to u\quad\text{in } L^r(T\times]0,\tau[)\text{ for all }\tau\in]0,\infty[\text{ and all } r\in[p,\frac{mp}{m-p}[. \tag{24}$$

Notice that to have *one* subsequence which works for *all* such τ and r, one can apply the usual diagonal sequence technique. By passing to a further subsequence if necessary, we can also assume that

$$u^{(j')}\to u\quad\text{point-wise almost everywhere on } T\times]0,\infty[\tag{25}$$

and

$$u^{(j')}\rightharpoonup u\quad\text{weakly* in } L^\infty(0,\infty;W^{1,p}(T,\mathbb{R}^{n+1})) \tag{26}$$

and finally, that

$$\partial_t u^{(j)}\rightharpoonup\partial_t u\quad\text{weakly in } L^2(T\times]0,\infty[). \tag{27}$$

Notice also, that by the growth condition (15) we imposed on ρ, the sequence $\{\sigma(Du^{(j)})\}_j$ is bounded in $L^\infty(0,\infty;L^{p'}(T))$. Therefore,

$$\sigma(Du^{(j')})\rightharpoonup w\quad\text{weakly* in } L^\infty(0,\infty;L^{p'}(T)) \tag{28}$$

for a further subsequence.

For a test function $\phi(x,t) = \xi(t)\eta(x)$, with $\xi \in C_0^\infty(]0,\infty[)$ and $\eta \in \mathrm{span}(w_1,\ldots,w_l)$, we infer from (20) that (as soon as $j' > l$)

$$\begin{array}{ccccccc} \displaystyle\int_0^\infty\int_T \partial_t u^{(j')}\phi & + & \displaystyle\int_0^\infty\int_T \sigma(Du^{(j')}) : D\phi & + & \displaystyle\int_0^\infty\int_T k\chi'(|u^{(j')}|)\frac{u^{(j')}}{|u^{(j')}|}\phi & = 0 \\ \downarrow & & \downarrow & & \downarrow \text{ as } j' \to \infty & \\ \displaystyle\int_0^\infty\int_T \partial_t u\phi & + & \displaystyle\int_0^\infty\int_T w : D\phi & + & \displaystyle\int_0^\infty\int_T k\chi'(|u|)\frac{u}{|u|}\phi & = 0. \end{array} \tag{29}$$

To pass to the limit in the first and the second term of (29), we used (27) and (28). For the third term, we used (25) and Lebesgue's dominated convergence theorem. By density of the linear span of test functions we used, the resulting equation in (29) follows also for arbitrary $\phi \in C_0^\infty(T\times]0,\infty[)$.

In order to identify w we now want to apply Minty's lemma for monotone operators. (see, e.g., [42, "decisive monotonicity trick"]). To do this, we first observe, that by (22), for $j' \to \infty$, and by using (24), (25) and (27)

$$\begin{aligned} \int_0^\tau\int_T -\sigma(Du^{(j')}) : Du^{(j')} &= \int_0^\tau\int_T \partial_t u^{(j')} u^{(j')} + \int_0^\tau\int_T k\chi'(|u^{(j')}|)\frac{u^{(j')}}{|u^{(j')}|}u^{(j')} \\ &\to \int_0^\tau\int_T \partial_t u u + \int_0^\tau\int_T k\chi'(|u|)\frac{u}{|u|}u. \end{aligned} \tag{30}$$

On the other hand, by (29), we have

$$\int_0^\tau\int_T -w : Du = \int_0^\tau\int_T u\partial_t u + \int_0^\tau\int_T k\chi'(|u|)\frac{u}{|u|}u \tag{31}$$

We now apply Minty's lemma for the monotone operator

$$\begin{aligned} A : L^p(0,\tau;W^{1,p}(T)) &\to L^{p'}(0,\tau;W^{-1,p'}(T)) \\ u &\mapsto \Big(\phi \mapsto \int_0^\tau\int_T \sigma(Du) : D\phi\Big). \end{aligned}$$

Now, as desired, from (26), (28), and the fact that the right-hand sides of (30) and (31) agree, it follows from Minty's lemma that $w = \sigma(Du)$. Hence, u is a weak solution of (18).

The energy inequality (19) for u follows from (23) as $j \to \infty$: We have

$$\liminf_{j\to\infty}\int_0^\tau \|\partial_t u^{(j)}\|^2_{L^2(T)} \geq \int_0^\tau \|\partial_t u\|^2_{L^2(T)}$$

by (27),

$$\liminf_{j\to\infty} E_k(u^{(j)}(\tau)) \geq E_k(u(\tau)) \qquad \text{for almost all } \tau$$

by (26), the convexity of ρ, and by (25). And finally, observe that

$$\lim_{j\to\infty} E_k(u_0^{(j)}) = E_k(u_0)$$

by (21), the Vitali convergence theorem and by (25). □

2.2. Passage to the limit I

We now want to pass to the limit in (18). Observe first, that by the energy inequality (19), the sequence $\{u_k\}$ we constructed is bounded in $L^\infty(0,\infty;W^{1,p}(T))$ and $\{\partial_t u_k\}$ is bounded in $L^2(0,\infty;L^2(T))$. By passing to a subsequence which we still denote by u_k, we may assume that

$$u_k \rightharpoonup u \quad \text{weakly* in } L^\infty(0,\infty;W^{1,p}(T)) \tag{32}$$

and

$$\partial_t u_k \rightharpoonup \partial_t u \quad \text{weakly in } L^2(T\times]0,\infty[). \tag{33}$$

Moreover, by Aubin's Lemma, we can extract a further subsequence such that

$$u_k \to u \quad \text{in } L^r(T\times]0,\tau[) \text{ for all } \tau \in]0,\infty[\text{ and all } r \in [p, \frac{mp}{m-p}[, \tag{34}$$

and hence, for yet a further subsequence

$$u_k \to u \quad \text{point-wise almost everywhere on } T\times]0,\infty[. \tag{35}$$

Notice, that from the energy inequality (19) it follows, that $|u_k| \to 1$ almost everywhere. Combining this with (35), we conclude, that $|u| = 1$ almost everywhere on $T \times [0,\infty[$, i.e., u takes values in the sphere, as desired. This is, nonetheless, not sufficient to already pass to the limit in (18).

The following lemma is a maximum principle for the weak solutions u_k of (18).

Lemma 9. *Let u_k be a weak solution of* (18) *on* $[0,\tau[$. *Then,* $|u_k(t)| \leq 1$ *for all* $t \in [0,\tau[$, *if* $|u_k(0)| \leq 1$.

Proof. To ease notation, we write u in place of u_k. We use

$$R(u) := u - \frac{u}{|u|}\min\{1,|u|\}$$

as a test function in (18) and find for the three single terms:

$$\int_T R(u)\partial_t u = \partial_t \int_T \xi(|u|)$$

for

$$\xi(x) = \begin{cases} 0 & \text{if } 0 \leq x < 1 \\ \frac{1}{2}(x-1)^2 & \text{if } x \geq 1. \end{cases}$$

Therefore

$$\int_T R(u)\partial_t u = \partial_t \int_{\{|u|\geq 1\}} (|u|-1)^2 .$$

In the second term, we get

$$\int_T k\chi'(|u|)\frac{u}{|u|}R(u) \geq 0$$

since $\chi'(|u|) \geq 0$ for $|u| \geq 1$ and $R(u) = 0$ for $|u| \leq 1$. For the third term, we have

$$\int_T \sigma(Du) : DR(u) = \int_{\{|u|\geq 1\}} \sigma(Du) : Du(1 - \frac{1}{|u|}) + \int_{\{|u|\geq 1\}} \rho'(|Du|)\frac{|u\,Du|^2}{|Du||u|^3} .$$

Observe, that the last term in the previous expression is non-negative. Hence, summarizing, we obtain

$$\frac{1}{2}\frac{\partial}{\partial t}\int_{\{|u|\geq 1\}}(|u|-1)^2+\int_{\{|u|\geq 1\}}\sigma(Du):Du(1-\frac{1}{|u|})\leq 0.$$

Since the second term is non-negative, the derivative in the first term is non-positive, and therefore the lemma follows. □

By (31), we have for arbitrary $\tau > 0$

$$\int_0^\tau\int_T \partial_t u_k u_k+\int_0^\tau\int_T \sigma(Du_k):Du_k+k\int_0^\tau\int_T \chi'(|u_k|)\frac{u_k}{|u_k|}u_k=0.$$

The energy inequality (23) and the growth condition (15) imply, that the first two terms are, for fixed τ, uniformly bounded in k. In view of Lemma 9, and the fact, that χ' vanishes on $[0,\frac{1}{2}]$ we obtain

$$\int_0^\tau\int_T \left|k\chi'(|u_k|)\frac{u_k}{|u_k|}\right|\leq -2\int_0^\infty\int_T k\chi'(|u_k|)\frac{u_k}{|u_k|}u_k\leq C$$

for a constant C which does not depend on k.

We conclude, that the right-hand side in

$$\partial_t u_k-\operatorname{div}\sigma(Du_k)=-k\chi'(|u_k|)\frac{u_k}{|u_k|}\tag{36}$$

is a bounded sequence in $L^1(T\times[0,\tau])$. This fact suggests now to consider the Young measure generated by the sequence $u_k\otimes Du_k$. Observe, that u_k converges almost everywhere, an therefore, if Du_k generates the Young measure ν on the space $\mathbb{M}^{(n+1)\times m}$ of $(n+1)\times m$ matrices, then $u_k\otimes Du_k$ generates the Young measure $\delta_u\otimes\nu$ (see, e.g., [27]). We use the next section to collect some relevant facts about the Young measure ν.

2.3. The Young measure generated by Du_k

Lemma 10. *Let ν be the Young measure generated by the the gradients Du_k of the solutions of* (18) *constructed in Lemma* 8. *Then*

(i) $\nu_{(x,t)}$ *is a probability Young measure for almost every* $(x,t)\in T\times[0,\tau]$.
(ii) $\nu_{(x,t)}$ *satisfies* $Du(x,t)=\langle\nu_{(x,t)},\mathrm{id}\rangle$ *for almost every* $(x,t)\in T\times[0,\tau]$.

Proof. (i) follows directly from the fact, that Du_k is bounded in $L^1(T\times[0,\tau])$ (see, e.g., [1] or [27]).

(ii): Du_k converges weakly to Du in $L^p(T\times[0,\tau])$. On the other hand, since Du_k is an equiintegrable sequence, we have that Du_k converges weakly to $\langle\nu_{(x,t)},\mathrm{id}\rangle$ in $L^1(T\times[0,\tau])$. Therefore, $Du(x,t)=\langle\nu_{(x,t)},\mathrm{id}\rangle$. □

The final ingredient we need is a "div-curl inequality":

$$\int_0^\tau\int_T\int_{\mathbb{M}^{(n+1)\times m}}(\sigma(\lambda)-\sigma(Du)):(\lambda-Du)d\nu_{(x,t)}(\lambda)\,dx\,dt\leq 0.\tag{37}$$

This inequality appears first in [8] for the Young measure ν generated by sequence of the gradients of solutions of a stationary non-linear elliptic system of p-Laplace type with right-hand side bounded in L^1. The class of operators considered there includes in particular our case $\sigma(Du)$. Due to the fact that, as we have seen, the right-hand side in (36) is bounded in L^1, the proof in [8] carries over to the parabolic case.

2.4. Passage to the limit II

The aim is now, to pass to the limit in the penalized equation (18). The problem in the operator term is, that ρ need not be strictly convex, and hence σ is not necessarily strictly monotone, or even uniformly monotone like in the case of the p-Laplace operator. Nonetheless, if ρ is not strictly convex in a region, it must be affine there, and it should therefore nevertheless be possible to pass to the limit there. It turns out that Young measures provide the adequate tool to go through with this program.

We start with some notation.

We consider the limit u of u_k as in (32)–(35). Let (x,t) be a fixed point in $T \times [0,\tau]$. Then, let

$$L := \{(\lambda, \rho(|Du(x,t)|) + \sigma(Du(x,t)) : (\lambda - Du(x,t))) \mid \lambda \in \mathbb{M}^{(n+1)\times m}\} \subset M^{(n+1)\times m} \times \mathbb{R}$$

denote the supporting hyperplane to the graph of $\rho(|\lambda|)$ in $(Du(x,t), \rho(|Du(x,t)|))$ (see Figure 2). Moreover, let

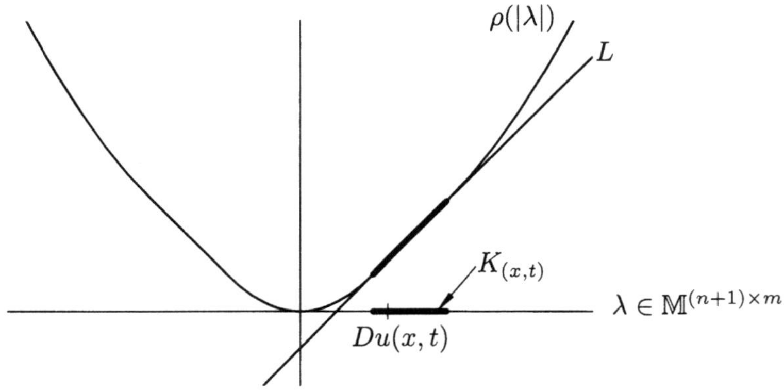

FIGURE 2. Graph of ρ

$$K_{(x,t)} = \{\lambda \in \mathbb{M}^{(n+1)\times m} \mid \rho(|\lambda|) = \rho(|Du(x,t)|) + \sigma(Du(x,t)) : (\lambda - Du(x,t))\}$$

denote the set, where ρ agrees with the supporting hyperplane L (see Figure 2).

Lemma 11. *For almost all $(x,t) \in T \times [0,\tau]$, the Young measure $\nu_{(x,t)}$ is supported in the set $K_{(x,t)}$, i.e., $\operatorname{spt}(\nu_{(x,t)}) \subset K_{(x,t)}$.*

Proof. By (37) the integrand $(\sigma(\lambda) - \sigma(Du(x,t))) : (\lambda - Du(x,t))$ (which is non-negative everywhere because of the monotonicity of σ) must vanish as a function of λ on the support of the measure $\nu_{(x,t)}$. This is true with the possible exception of a set N of $(x,t) \in T \times [0,\tau]$ of measure zero. Let us fix $\lambda \in \mathrm{spt}(\nu_{(x,t)})$ for $(x,t) \notin N$, then

$$(1-\alpha)(\sigma(Du(x,t)) - \sigma(\lambda)) : (Du(x,t) - \lambda) = 0 \quad \text{for all } \alpha \in [0,1]. \tag{38}$$

On the other hand, again by the monotonicity of σ, we have

$$0 \leq (1-\alpha)(\sigma(Du(x,t)+\alpha(\lambda - Du(x,t))) - \sigma(\lambda)) : (Du(x,t)-\lambda) \quad \text{for all } \alpha \in [0,1]. \tag{39}$$

Subtracting (38) from (39) we obtain

$$0 \leq (1-\alpha)(\sigma(Du(x,t)+\alpha(\lambda - Du(x,t))) - \sigma(Du(x,t))) : (Du(x,t)-\lambda) \quad \text{for all } \alpha \in [0,1]. \tag{40}$$

But by monotonicity, also the opposite inequality holds true:

$$0 \geq (1-\alpha)(\sigma(Du(x,t)+\alpha(\lambda - Du(x,t))) - \sigma(Du(x,t))) : (Du(x,t)-\lambda) \quad \text{for all } \alpha \in [0,1]. \tag{41}$$

Therefore, still for the same $\lambda \in \mathrm{spt}(\nu_{(x,t)})$, by (40) and (41), we have

$$0 = (1-\alpha)(\sigma(Du(x,t)+\alpha(\lambda - Du(x,t))) - \sigma(Du(x,t))) : (Du(x,t)-\lambda) \quad \text{for all } \alpha \in [0,1]. \tag{42}$$

Using (42), we conclude

$$\begin{aligned} \rho(|\lambda|) &= \rho(|Du(x,t)|) + \int_0^1 \sigma(Du(x,t)+\alpha(\lambda - Du(x,t)) : (\lambda - Du(x,t))d\alpha = \\ &= \rho(|Du(x,t)|) + \sigma(Du(x,t)) : (\lambda - Du(x,t)) \end{aligned}$$

and hence, $\lambda \in K_{(x,t)}$, as claimed. □

Lemma 12. (i) *For each $\lambda \in K_{(x,t)}$ there holds $\sigma(\lambda) = \sigma(Du(x,t))$.*
(ii) *For almost all $(x,t) \in T \times [0,\tau]$ the following is true:*

$$\int_{M^{(n+1)\times m}} \sigma(\lambda) d\nu_{(x,t)}(\lambda) = \sigma(Du(x,t)). \tag{43}$$

Proof. (i) follows the fact that ρ is continuously differentiable.

(ii): As we have seen in Lemma 11, the support of the measure $\nu_{(x,t)}$ is contained in the set $K_{(x,t)}$. Therefore, in (43), we only need to integrate over $K_{(x,t)} \subset M^{(n+1)\times m}$. But there, as stated in (i), $\sigma(\lambda) = \sigma(Du(x,t))$. Hence, the claim follows from the fact that, by Lemma 10(i), $\nu_{(x,t)}$ has mass 1. □

Actually, from (ii) in the previous lemma and the fundamental theorem on Young measures (see, e.g., [1]) it now already follows, that $\sigma(Du_k) \rightharpoonup \sigma(Du)$ in $L^1(T \times [0,\tau]$. However, as we will see in the next lemma, this convergence is even strong.

Lemma 13. *Let u_k be the sequence solutions of* (18) *constructed in Lemma* 8, *u the limit of this sequence in* (32), *and*

$$g_k(x,t) := |\sigma(Du_k(x,t)) - \sigma(Du(x,t))|$$

Then, $g_k \to 0$ in $L^1(T \times [0,\tau])$ for all $\tau > 0$.

Proof. Observe, that, since $\{Du_k\}$ is bounded in $L^\infty(0,\infty; L^p(T))$, by (15) $\sigma(Du_k)$ is bounded in $L^\infty(0,\tau; L^{p'}(T))$. Therefore, g_k is an equiintegrable sequence on $T \times [0,\tau]$. Hence, by the fundamental theorem of Young measures (see, e.g., [1]), for a suitable (not relabeled) subsequence, $g_k \rightharpoonup \bar{g}$ in $L^1(T \times [0,\tau])$ where

$$\bar{g}(x,t) = \int_{\mathbb{M}^{(n+1)\times m}} |\sigma(\lambda) - \sigma(Du(x,t))| d\nu_{(x,t)} = 0$$

Here, we have used that by Lemma 11, we only have to integrate over $K_{(x,t)}$ and that there, by Lemma 12(i), $\sigma(\lambda) = \sigma(Du(x,t))$. Since $g_k \geq 0$, the convergence is strong. By the usual diagonal sequence argument, we can find a subsequence which converges strongly in $L^1_{\text{loc}}(T \times [0,\infty])$. □

This allows us now to finally prove the existence theorem for the heat flow of the energy E:

Theorem 14. *If the convex C^1-function ρ satisfies the coercivity and growth conditions* (14) *and* (15), *then there exists a global weak solution of* (16) *for arbitrary initial data $u_0 \in W^{1,p}(T, S^n)$.*

Proof. It follows from Lemma 13, that

$$\sigma(Du_k) \to \sigma(Du) \quad \text{in } L^1_{\text{loc}}(T \times [0,\infty]) \tag{44}$$

By taking the wedge product of (18) with u_k, we find that

$$0 = \partial_t u_k \wedge u_k - \operatorname{div}(\rho'(|Du_k|)\frac{Du_k}{|Du_k|} \wedge u_k). \tag{45}$$

Passing to the limit in the weak form of (45), and using (33), (34) and (44) yields

$$0 = \partial_t u \wedge u - \operatorname{div}(\rho'(|Du|)\frac{Du}{|Du|} \wedge u). \tag{46}$$

As stated at the beginning of Section 2.2, $|u| = 1$ almost everywhere on $T \times [0,\infty[$. Therefore,

$$(\partial_t u - \operatorname{div}(\sigma(Du)) - u\sigma(Du) : Du)u\psi = 0 \tag{47}$$

is an identity for arbitrary $\psi \in C_0^\infty(T \times [0,\infty])$.

Let $\phi \in C_0^\infty(T \times [0,\infty], \mathbb{R}^{n+1})$. We test (46) by $u \wedge \phi$ and use $\psi = u \cdot \phi$ in (47) (observe, that these are admissible test-functions) and subtract the resulting equations from each another. Using the identity

$$\phi = u(u \cdot \phi) - u \wedge (u \wedge \phi)$$

this yields the weak form of (16). □

References

[1] J.M. Ball: A version of the fundamental theorem for Young measures. In: Partial differential equations and continuum models of phase transitions: Proceedings of an NSF-CNRS joint seminar. Springer, 1989

[2] F. Bethuel: The approximation problem for Sobolev maps between manifolds. Acta Math. **167**, 153–206 (1991)

[3] F. Bethuel, X. Zheng: Density of smooth functions between two manifolds in Sobolev spaces. J. Funct. Anal. **80**, 60–75 (1988)

[4] K.-C. Chang, W.-Y. Ding, R. Ye, R.: Finite-time blow-up of the heat flow of harmonic maps from surfaces. J. Differ. Geom. **36**, No. 2, 507–515 (1992)

[5] Y. Chen: The weak solutions to the evolution problem of harmonic maps. Math. Z. **201**, 69–74 (1989)

[6] Y. Chen, M.-C. Hong, N. Hungerbühler: Heat flow of p-harmonic maps with values into spheres. Math. Z. **215**, 25–35 (1994)

[7] P. Courilleau, F. Demengel: Heat flow for p-harmonic maps with values in the circle. Nonlinear Anal. **41**, no. 5-6, Ser. A: Theory Methods, 689–700 (2000)

[8] G. Dolzmann, N. Hungerbühler, S. Müller: Non-linear elliptic systems with measure-valued right-hand side. Math. Z. **226**, 545–574 (1997)

[9] F. Duzaar, M. Fuchs: On removable singularities of p-harmonic maps. Analyse non linéaire, **7**, No. 5 385–405 (1990)

[10] F. Duzaar, M. Fuchs: Existence and regularity of functions which minimize certain energies in homotopy classes of mappings. Asymptotic Analysis **5**, 129–144 (1991)

[11] J. Eells, L. Lemaire: Another report on harmonic maps. Bull. London Math. Soc. **20**, 385–524 (1988)

[12] J. Eells, J.H. Sampson: Harmonic mappings of Riemannian manifolds. Am. J. Math. **86**, 109–169 (1964)

[13] J. Eells, J.C. Wood: Restrictions on harmonic maps of surfaces. Topology **15**, 263–266 (1976)

[14] A. Fardoun, R. Regbaoui: Équation de la chaleur pour les applications p-harmoniques entre variétés riemanniennes compactes. (French) C. R. Acad. Sci. Paris Sér. I Math. **333**, no. 11, 979–984 (2001)

[15] A. Fardoun, R. Regbaoui: Heat flow for p-harmonic map with small initial data. To appear in Calc. Var. Partial Differential Equations

[16] A. Fardoun, R. Regbaoui: Heat flow for p-harmonic maps between compact Riemannian manifolds. To appear in Indiana Univ. Math. J.

[17] H. Federer: Geometric measure theory. Springer, 1969

[18] M. Fuchs: Some regularity theorems for mappings which are stationary points of the p-energy functional. Analysis **9**, 127–143 (1989)

[19] M. Fuchs: p-harmonic obstacle problems. Part III: Boundary regularity. Annali Mat. Pura Applicata **156**, 159–180 (1990)

[20] R.S. Hamilton: Harmonic maps of manifolds with boundary. Lect. Notes Math. **471**, Springer, Berlin, 1975

[21] R. Hardt, F.H. Lin: Mappings minimizing the L^p norm of the gradient. Comm. Pure and appl. Math **15**, 555–588 (1987)

[22] N. Hungerbühler: p-harmonic flow. Diss. Math. Wiss. ETH Zürich, Nr. 10740, 1994. Ref.: Michael Struwe; Korref.: Jürgen Moser, Zürich, 1994

[23] N. Hungerbühler: Global weak solutions of the p-harmonic flow into homogeneous spaces. Indiana Univ. Math. J. **45/1**, 275–288 (1996)

[24] N. Hungerbühler: Compactness properties of the p-harmonic flow into homogeneous spaces. Nonlinear Anal. **28/5**, 793–798 (1997)

[25] N. Hungerbühler: Non-uniqueness for the p-harmonic flow. Canad. Math. Bull. **40/2**, 174–182 (1997)

[26] N. Hungerbühler: m-harmonic flow. Ann. Scuola Norm. Sup. Pisa Cl. Sci. (4) **XXIV**, **4**, 593–631 (1997)

[27] N. Hungerbühler: A refinement of Ball's Theorem on Young measures. New York J. Math. **3**, 48–53 (1997)

[28] N. Hungerbühler: Young measures and nonlinear PDEs. Habilitationsschrift ETH Zürich, 2000

[29] N. Hungerbühler: Quasilinear parabolic systems in divergence form with weak monotonicity. Duke Math. J. **107/3**, 497–520 (2001)

[30] L. Lemaire: Applications harmoniques de surfaces riemannienne. J. Diff. Geom. **13**, 51–78 (1978)

[31] X.G. Liu, S.H. Li: The p-harmonic heat flow with potential into homogeneous spaces. Acta Math. Sin. (Engl. Ser.) **18**, no. 1, 21–26 (2002)

[32] X.G. Liu: A note on heat flow of p-harmonic mappings. (Chinese) Kexue Tongbao (Chinese) **42**, no. 1, 15–18 (1997)

[33] X.G. Liu: A remark on p-harmonic heat flows. Chinese Sci. Bull. **42**, no. 6, 441–444 (1997)

[34] M. Misawa: On regularity for heat flows for p-harmonic maps. Proceedings of the Korea-Japan Partial Differential Equations Conference (Taejon, 1996), 14 pp., Lecture Notes Ser., **39**, Seoul Nat. Univ., Seoul, 1997

[35] M. Misawa: Existence and regularity results for the gradient flow for p-harmonic maps. Electron. J. Differential Equations, No. **36**, 17 pp. (1998)

[36] M. Misawa: Existence and regularity results for the gradient flow for p-harmonic maps. Regularity, blowup and related properties of solutions to nonlinear evolution equations (Japanese). (Kyoto, 1997). Sūrikaisekikenkyūsho Kōkyūroku No. **1045**, 57–72 (1998)

[37] M. Misawa: On the p-harmonic flow into spheres in the singular case. Nonlinear Anal. **50**, no. 4, Ser. A: Theory Methods, 485–494 (2002)

[38] J. Sacks, K. Uhlenbeck: The existence of minimal immersions of 2-spheres. Ann. of Math. **113**, 1–24 (1981)

[39] R.S. Schoen, K. Uhlenbeck: Approximation theorems for Sobolev mappings, preprint

[40] P. Strzelecki: Regularity of p-harmonic maps from the p-dimensional ball into a sphere. Manuscr. Math. **82**, No.3–4, 407–415 (1994)

[41] P. Strzelecki: Stationary p-harmonic maps into spheres. Janeczko, Stanislaw (ed.) et al., Singularities and differential equations. Proceedings of a symposium, Warsaw, Poland. Warsaw: Polish Academy of Sciences, Inst. of Mathematics, Banach Cent. Publ. **33**, 383–393 (1996)

[42] E. Zeidler: Nonlinear functional analysis and its applications,Vol. II/B: Nonlinear Monotone Operators. New York: Springer, 1990

[43] C.Q. Zhou: Compactness properties of heat flows for weakly p-harmonic maps. (Chinese) Acta Math. Sinica **41**, no. 2, 327–336 (1998)

[44] C.Q. Zhou, Y.D. Wang: Existence and nonuniqueness for the flow of p-harmonic maps. (Chinese) Acta Math. Sinica **41**, no. 3, 511–516 (1998)

Norbert Hungerbühler
Department of Mathematics
University of Fribourg
Pérolles
CH-1700 Fribourg, Switzerland

Progress in Nonlinear Differential Equations
and Their Applications, Vol. 59, 67–80

Singularity Models for the Ricci Flow: An Introductory Survey

Dan Knopf

Abstract. At the time this article is being written (May 2003) much of the mathematical world is waiting with intense interest to see the results of Grisha Perelman's effort [22, 23] to resolve William Thurston's Geometrization Conjecture [28] for closed 3-manifolds by completing the program [16] begun by Richard Hamilton. It is still too early to give an accurate and fair assessment of the full impact of Perelman's work. But in order to aid the many mathematicians who may be inspired by that work to look more closely at the Ricci flow, this does seem like an appropriate time to write a brief and purely expository introduction to the topic, intended for the non-expert. Readers desiring more information are encouraged to read the more advanced survey articles [6] and [7], as well as to consult Hamilton's and Perelman's original papers.

1. Heuristics

There are two heuristic principles which are useful to keep in mind when one first studies the (unnormalized) Ricci flow, wherein one starts with a smooth Riemannian manifold $(\mathcal{M}^n, g_0)$ and evolves it by the equation

$$\frac{\partial}{\partial t} g = -2\,\mathrm{Rc}, \tag{1.1}$$

where Rc denotes the Ricci tensor of the metric g. The first principle is that equation (1.1) is morally the heat equation for a Riemannian metric. The best way to see this is by writing the right-hand side of equation (1.1) in harmonic coordinates. Recall that a coordinate chart $\varphi : \mathcal{U} \to \mathbb{R}^n$ defined in a neighborhood $\mathcal{U}$ of a point x_0 on a smooth Riemannian manifold $(\mathcal{M}^n, g)$ is called *harmonic* if the coordinate functions $\{x^k : 1 \leq k \leq n\}$ it induces are harmonic throughout $\mathcal{U}$:

$$\Delta x^k = 0.$$

It follows from standard existence and regularity theory for elliptic PDE that a harmonic coordinate chart exists in a sufficiently small neighborhood $\mathcal{U}$ of any $x_0 \in$

Received by the editors Version 071103.
Partially supported by NSF grant DMS-0328233.

$\mathcal{M}^n$, and moreover that the metric enjoys optimal regularity in such coordinates. The feature of harmonic coordinates relevant to the Ricci flow is that the identity

$$0 = -\Delta x^k = -g^{ij}\left(\frac{\partial^2}{\partial x^i \partial x^j} - \Gamma_{ij}^{\ell}\frac{\partial}{\partial x^{\ell}}\right)x^k = g^{ij}\Gamma_{ij}^k$$

holds throughout $\mathcal{U}$. Using this, it is not hard to see that

$$-2R_{ij} = -2R_{mij}^{m} = g^{k\ell}\frac{\partial^2}{\partial x^k \partial x^{\ell}}\left(g_{ij}\right) + Q_{ij}\left(g^{-1}, \partial g\right), \tag{1.2}$$

where Q is quadratic in the inverse and first derivatives of g. In other words, the highest-order derivative term in $-2R_{ij}$ appears in a harmonic chart to be the Laplacian of the component g_{ij} of the metric, regarded as a scalar function in that chart.

There are at least two reasons why this principle is heuristic and not entirely rigorous. One is that equation (1.2) is not tensorial; indeed, all covariant derivatives of the metric vanish identically. Another reason is that when one evolves a metric by (1.1), coordinates which are harmonic at time t cannot be expected to be so at time $t + \varepsilon$, for any $\varepsilon > 0$.

Nonetheless, much of what this heuristic principle suggests is almost true. Equation (1.2) is a quasilinear parabolic equation. (It is quasilinear because the inverse of the unknown function $g(t)$ multiplies the highest-order derivatives in the equation.) In fact, the Ricci flow is quasilinear and almost parabolic. Its failure to be strictly parabolic stems from the fact that the Ricci flow is defined entirely in terms of natural geometric quantities, and hence is invariant under the full diffeomorphism group. This invariance is a great advantage from the standpoint of geometry. Fortunately, it is only a minor inconvenience from the standpoint of PDE, because Dennis DeTurck showed [10, 11] that the flow is equivalent to a strictly parabolic quasilinear equation, so that short-time existence and uniqueness follow from standard theory.

The heuristic of thinking of the Ricci flow as a heat equation is useful in another important sense. The heat equation seeks to regularize its initial data, so one expects equation (1.1) to improve the metric, at least for a short time. In fact, initial bounds on the curvature of a metric induce subsequent *a priori* bounds on all derivatives of the curvature. These derivative bounds were derived in [3], [2], and [25, 26]; most known proofs use the maximum principle for parabolic equations in a familiar technique pioneered by Bernstein.

On a more intuitive level, the heat equation heuristic leads one to expect that the Ricci flow will mimic the diffusion properties of the heat equation, and thus will try to make a metric more homogeneous and isotropic. As we will see below, this expectation too is often justified.

The second heuristic principle is that we should expect the Ricci flow to develop singularities; in particular, it tells us that the first principle can be misleading if we take it too literally. One arrives at the second heuristic from the viewpoint of geometry when one observes that the normalized Ricci flow (defined in (2.1) below) can converge only to an Einstein metric. Since most Riemannian manifolds of dimension $n > 2$ are not Einstein, one should expect something to go wrong. The second principle is also supported by the viewpoint of analysis. Indeed, equation (1.1) implies that the scalar curvature evolves by

$$\frac{\partial}{\partial t} R = \Delta R + 2\left|\mathrm{Rc}\right|^2 . \tag{1.3}$$

This is a reaction-diffusion equation: the reaction term $2\left|\mathrm{Rc}\right|^2$ may be regarded as fighting against the diffusion term ΔR. By a standard estimate, equation (1.3) implies that

$$\frac{\partial}{\partial t} R \geq \Delta R + \frac{2}{n} R^2 .$$

By the parabolic maximum principle, one can then compare solutions of (1.3) with solutions of the ODE

$$\frac{d\rho}{dt} = \frac{2}{n}\rho^2$$

obtained by ignoring the diffusion term in (1.3). One concludes that

$$R_{\min}(t) \geq \frac{1}{R_{\min}^{-1}(t_0) - \frac{2}{n}(t - t_0)}$$

for all $t \geq t_0$ that the solution $g(t)$ exists. Hence if the scalar curvature ever becomes everywhere positive, a finite-time singularity is inevitable.

The second heuristic principle suggests correctly that a positive resolution of the Geometrization Conjecture depends on developing an adequate understanding of singularities. One must learn what the behaviour of a solution to the Ricci flow that becomes singular reveals about the topology of the underlying manifold. The reason why current research into geometrization concentrates on the analysis of singularities is because any solutions which remain nonsingular behave very nicely. Indeed, these are the solutions for which the diffusion term dominates, in accord with the first heuristic principle. Hence our informal survey begins with such solutions.

2. Nonsingular solutions

By rescaling space and time, one can modify the Ricci flow so that it preserves volume. This leads to the equation

$$\frac{\partial}{\partial t} g = -2\,\mathrm{Rc} + \frac{2r}{n} g, \tag{2.1}$$

where

$$r(t) = \frac{\int_{\mathcal{M}^n} R\, d\mu}{\int_{\mathcal{M}^n} d\mu}$$

denotes the average scalar curvature. Equation (1.1) (the unnormalized Ricci flow) is in a sense the more natural PDE, but equation (2.1) (the normalized Ricci flow) can be more convenient for taking limits and for establishing certain convergence properties.

For example, in proving Hamilton's seminal result [14] that any manifold which admits a metric of positive Ricci curvature is in fact a space form, one first studies the solution of equation (1.1). One shows that the solution exists on a maximal time interval $0 \leq t < T < \infty$, that the curvature becomes unbounded as $t \nearrow T$, and that the metric becomes nearly Einstein at points where the curvature is large. Then one derives a crucial estimate on the gradient of the scalar curvature. Together with the observation that the diameter of the solution is uniformly bounded, this allows one to show that the ratio $R_{\max}/R_{\min}$ approaches 1 as the singularity time is approached. Then one rescales space and time, converting the solution $g(t)$ of equation (1.1) into a solution $\bar{g}(\bar{t})$ of equation (2.1). In the final step, one shows that $\bar{g}(\bar{t})$ exists for all time and converges exponentially in every C^k norm to a metric of constant sectional curvature.

One way of understanding the role of the normalized flow in this proof is as follows: because the minimum curvature of the original solution approaches the maximum curvature as the maximum becomes large, the average curvature itself becomes large. Intuitively this means that when one rescales, the dilation term $2r/n$ is large enough to keep the solution nonsingular. So in this case, the diffusion effect suggested by the first heuristic principle wins: it tames the singularity one expects from the second heuristic principle. On the other hand, if the curvature of a solution of (2.1) ever becomes large on a set of small volume, one would not expect the dilation term to be adequate to avoid a singularity.

One says a solution $(\mathcal{M}^3, g(t))$ of the normalized flow (2.1) on a compact 3-manifold is *nonsingular* if it exists for all positive time and satisfies a uniform curvature bound

$$\sup_{\mathcal{M}^3 \times [0,\infty)} |\mathrm{Rm}| \leq C < \infty.$$

In this case, $\mathcal{M}^3$ is geometrizable. This result was proved in [18]. The proof uses the Gromov-type convergence results in [17] as well as the advanced tensor maximum principle of [15], which lets one compare a tensor evolving by a PDE with a solution to a system of ODE.

The results in [18] classify nonsingular solutions. The behaviours one observes for such solutions turn out to be instructive when one later studies singular solutions. Hence it will be useful to recall some aspects of that classification here.

If

$$\lim_{t \to \infty} \left(\sup_{x \in \mathcal{M}^3} \mathrm{inj}(x,t) \right) = 0, \tag{2.2}$$

one says the solution exhibits *collapse with bounded curvature*. In this case, results of Cheeger-Gromov imply that $\mathcal{M}^3$ admits an $\mathcal{F}$-structure, hence is a graph manifold, hence can be decomposed into a union of Seifert fiber-space pieces, all of which are known to be geometrizable. (An excellent survey is [24].) In case (2.2) does not hold, one can find $\varepsilon > 0$ and sequences of points $x_j \in \mathcal{M}^3$ and times

$t_j \to \infty$ such that

$$\inf_{j \in \mathbb{N}} (\text{inj}(x_j, t_j)) \geq \varepsilon. \tag{2.3}$$

Then there exist diffeomorphisms $\varphi_j : \mathcal{M}^3 \to \mathcal{M}^3$ such that the pointed sequence of solutions

$$(\mathcal{M}^3, g_j(t), x_j)$$

defined by

$$g_j(t) = (\varphi_j^* g)(t_j + t)$$

converges locally smoothly in the pointed category to a limit solution

$$(\mathcal{M}_\infty^3, g_\infty(t), x_\infty).$$

If $\mathcal{M}_\infty^3$ is compact, it is necessarily diffeomorphic to $\mathcal{M}^3$. In this case, results in [18] prove that g_∞ is a metric of constant sectional curvature K, hence that $\mathcal{M}^3$ is a space form. Moreover, one can in some cases prove a stronger convergence than was stated above. Indeed, if $K > 0$, then there exists t_j large enough so that the Ricci curvature of $g(t_j)$ is strictly positive; it then follows from Hamilton's original result [14] that the original solution converges exponentially fast to a space form. (In other words, one needed neither to modify by diffeomorphisms nor to restrict to a sequence of times.) If $K = 0$, the same statement holds by [13]. It is conjectured but not yet proven that a similar statement holds for the case $K < 0$.

The remaining possibility is that $\mathcal{M}_\infty^3$ is noncompact, which is in a sense the most interesting case. A particular consequence of the results proved in [18] is that there exists a finite collection of finite-volume noncompact hyperbolic manifolds $(\mathcal{H}_i, h_i)$ with the following properties. For each $k \in \mathbb{N}$, there exist truncations $\mathcal{H}_{i,k}$ of $\mathcal{H}_i$ along constant mean curvature tori of area less than $1/k$, a time t_k, and embeddings $\psi_{i,k}(t) : \mathcal{H}_{i,k} \hookrightarrow \mathcal{M}_\infty$ defined for all $t \geq t_k$ such that the images $\psi_{i,k}(t)(\mathcal{H}_{i,k})$ are mutually disjoint submanifolds for all $t \geq t_k$ and

$$\sup_{t \geq t_k} \left\| \psi_{i,k}^*(t)(g(t)) - h_i \right\|_{C^k(\mathcal{H}_{i,k})} \leq 1/k$$

in the norm of C^k convergence on compact subsets of $\mathcal{H}_{i,k}$. From the perspective of topology, the most important aspect of this case is that the fundamental group of each $\mathcal{H}_i$ injects into $\mathcal{M}_\infty^3$ under the maps $\psi_{i,k}$. This proves that $\mathcal{M}_\infty^3$ is Haken, hence is geometrizable by Thurston's result [28]. In fact, one can show directly that the injectivity radius at any point in the complement $\mathcal{M}_\infty \backslash \cup_i \psi_{i,k}(t)(\mathcal{H}_{i,k})$ of the hyperbolic pieces in the limit remains less than $1/k$ for all $t \geq t_k$. In particular, this implies that the complement is a manifold-with-boundary which admits an $\mathcal{F}$-structure, hence may be decomposed into a union of Seifert fiber-space pieces.

3. Singular solutions

Having seen that any manifold which admits a nonsingular solution of the Ricci flow is known to be geometrizable, one is led to ask the following question: What does a singularity of the flow reveal about the topology of the underlying manifold?

A powerful technique in geometric analysis for answering this question is to 'blow-up' the singularity, obtaining a limit flow whose properties should yield information about the geometry of the original manifold near the singularity just prior to its formation.

To see how this is done, consider a solution $(\mathcal{M}^n, g(t))$ of the unnormalized Ricci flow on a maximal time interval $0 \le t < T$. For simplicity, we shall discuss the case that $T < \infty$, although an analogous theory is available for singularities which form in infinite time. In the case of a finite-time singularity, it follows from short-time existence results for the flow that

$$\lim_{t \nearrow T} \left(\sup_{x \in \mathcal{M}^n} |\mathrm{Rm}(x,t)| \right) = \infty. \tag{3.1}$$

To develop intuition, consider what is perhaps the simplest example of a finite-time singularity. Let g_{can} denote the canonical metric of constant sectional curvature $K = 1$ on the sphere $\mathcal{S}^n$. Consider a 1-parameter family $g(t)$ of conformal metrics defined by

$$g(t) = r(t)^2 g_{\mathrm{can}}, \tag{3.2}$$

noting that $g(t)$ has constant sectional curvature $r(t)^{-2}$. Observe that $g(t)$ is a solution of the Ricci flow if and only if

$$2r\frac{dr}{dt} \cdot g_{\mathrm{can}} = \frac{\partial}{\partial t} g = -2\,\mathrm{Rc}[g] = -2\,\mathrm{Rc}[g_{\mathrm{can}}] = -2(n-1) \cdot g_{\mathrm{can}},$$

hence if and only if

$$r(t) = \sqrt{r_0^2 - 2(n-1)t} = \sqrt{2(n-1)}\sqrt{T-t}. \tag{3.3}$$

So if $g(t)$ solves the Ricci flow, it must become singular at $T = r_0^2/(2(n-1))$.

In general, one says that a finite-time singularity of $(\mathcal{M}^n, g(t))$ is of *Type I* or *fast-forming* if it occurs at the natural rate suggested by this example, hence if there exists $C < \infty$ such that

$$\sup_{\mathcal{M}^n \times [0,T)} |\mathrm{Rm}| \cdot (T-t) \le C < \infty.$$

On the other hand, one says the singularity is of *Type IIa* if

$$\sup_{\mathcal{M}^n \times [0,T)} |\mathrm{Rm}| \cdot (T-t) = \infty.$$

Solutions with this property are also called *slowly-forming singularities.* (This terminology may be somewhat nonintuitive initially, but is actually perfectly natural for reasons involving the sharpness of certain *a priori* estimates which guarantee short-time existence of the flow.)

It is common to study a singularity by the technique of *parabolic dilation.* One chooses sequences of points $x_j \in \mathcal{M}^n$, times $t_j \in [0,T)$ increasing monotonically to

T, and dilating factors $\lambda_j > 0$. Using these, one defines a sequence $(\mathcal{M}^n, g_j(t), x_j)$ of pointed solutions to the Ricci flow, where

$$g_j(t) = \lambda_j g\left(t_j + \frac{t}{\lambda_j}\right).$$

Notice that each solution $g_j(t)$ exists on the time interval $-\lambda_j t_j \leq t < \lambda_j (T - t_j)$.

One chooses the sequences $\{x_j\}$, $\{t_j\}$, and $\{\lambda_j\}$ in order to get a good limit. A flat limit would not be desirable because it would not reveal enough about the geometry of the original solution. (Intuitively, getting a flat limit means one has dilated too much.) On the other hand, one wants uniform bounds on the curvatures of the metrics $g_j(0)$ in order to take advantage of the derivative estimates described in Section 1. (Bounds on the curvature of a solution to the Ricci flow imply bounds on all derivatives of the curvature.) Accordingly, one usually chooses x_j and λ_j so that the curvature $|\mathrm{Rm}(x_j, t_j)|$ is comparable to $\sup_{x \in \mathcal{M}^n} |\mathrm{Rm}(x, t_j)|$ and so that

$$0 < c \leq \lambda_j^{-1} |\mathrm{Rm}(x_j, t_j)| \leq C < \infty. \tag{3.4}$$

Recalling (3.1), one chooses the sequence of times so that $\sup_{x \in \mathcal{M}^n} |\mathrm{Rm}(x, t_j)|$ is comparable to the maximum curvature over a sufficiently large interval of earlier times.

Having made suitable choices, one wants to prove that $(\mathcal{M}^n, g_j(t), x_j)$ converges locally smoothly in the pointed category to a limit solution $(\mathcal{M}^n_\infty, g_\infty(t), x_\infty)$ of the Ricci flow, called a *singularity model* (also called a 'final time limit flow' in the literature). As we shall see below, such solutions are typically very special. (For example, a singularity model formed from a Type IIa singularity will exist for all times $-\infty < t < \infty$, which helps explain why such singularities are called 'slowly forming'.) By developing an adequate understanding of these model solutions, one hopes to obtain local information about the geometry of the original sequence just prior to the formation of the singularity. In particular, the strategy for extracting topological information about a 3-manifold $\mathcal{M}^3$ from a solution $(\mathcal{M}^3, g(t))$ of the Ricci flow that becomes singular is to perform a geometric-topological surgery on $\mathcal{M}^3$ just prior to the singularity in such a way that the maximum curvature of the solution is reduced by an amount large enough to permit the flow to be continued on the piece or pieces that remain after the surgery. In the final step of this program, one needs to argue that only geometrically recognizable pieces will remain after finitely many surgeries.

The main difficulty in obtaining convergence to a singularity model is obtaining an adequate injectivity radius estimate for the sequence $(\mathcal{M}^n, g_j(t), x_j)$, namely an estimate of the form

$$\inf_{x \in \mathcal{M}^3} (\mathrm{inj}\,(x,t)) \geq \frac{c}{\sqrt{\sup_{x \in \mathcal{M}^3} |\mathrm{Rm}(x,t)|}}.$$

Until recently, such estimates had to be derived by *ad hoc* means. For example, in Section 23 of [16], Hamilton proved an isoperimetric inequality that implies an injectivity radius estimate for appropriately chosen sequences of dilations approaching a Type I singularity in dimension three. In Section 22 of the same paper, he proved an injectivity radius estimate for odd-dimensional solitons that is useful for dimension reduction. (See Section 5, below.) The paper [8] derives an injectivity estimate useful for forming noncompact limits of Type IIa singularities. One could cite other examples, but the recent work [22] of Perelman yields the most powerful and general estimate known to date, which enables one to construct singularity models under very general hypotheses.

4. Singularity models

By analogy with other geometric evolution equations (especially the mean curvature flow) one expects singularity models to have special properties. Indeed, such solutions are generally *ancient* (existing for $-\infty < t < \omega$) or *immortal* (existing for $\alpha \leq t < +\infty$) or even *eternal* (existing for $-\infty < t < +\infty$). Many singularity models also exhibit special symmetries or asymptotic symmetries. (See [9].)

An important class of singularity models is the set of *generalized fixed points.* Such solutions can be described in either of two equivalent ways. One says $g(t)$ is a *self-similar solution* of the Ricci flow if there exist scalars $\sigma(t)$ and diffeomorphisms ψ_t of $\mathcal{M}^n$ such that

$$g(t) = \sigma(t)\, \psi_t^* (g_0)\,. \tag{4.1}$$

Notice that a self-similar solution changes only by diffeomorphism and rescaling, hence may be regarded as a fixed point of the Ricci flow modulo diffeomorphisms. On the other hand, one says that a fixed Riemannian manifold $(\mathcal{M}^n, g_0)$ is a *Ricci soliton* if the identity

$$-2\,\mathrm{Rc}(g_0) = \mathcal{L}_X g_0 + 2\lambda g_0 \tag{4.2}$$

holds for some constant λ and some complete vector field X on $\mathcal{M}^n$. Notice that (4.2) is a coupled elliptic system for g_0 and X. Because a Ricci soliton is Einstein if the vector field X vanishes identically, any solution of (4.2) may be regarded as a generalization of an Einstein metric. One says a Ricci soliton is *shrinking, translating,* or *expanding* in the cases $\lambda < 0$, $\lambda = 0$, $\lambda > 0$ respectively.

It is not hard to show that there is a one-to-one relationship between self-similar solutions of the flow and Ricci solitons. Hence the two concepts are commonly regarded as equivalent.

The best-known Ricci soliton is the *cigar* $(\mathbb{R}^2, g_\Sigma)$ found by Hamilton. This is the complete metric

$$g_\Sigma = \frac{dx^2 + dy^2}{1 + x^2 + y^2}$$

on $\mathbb{R}^2$ of positive scalar curvature

$$R_\Sigma = \frac{4}{1 + x^2 + y^2}.$$

It is not hard to show that the curvature of the cigar decays like e^{-2s}, where $s = \operatorname{arcsinh} \sqrt{x^2 + y^2}$ is the metric distance from the origin. (The cigar is actually one representative of a family of Kähler-Ricci metrics that exist on $\mathbb{C}^{2m}$ and certain other complex manifolds. These are studied in [4, 5] and [12].) The reason why the cigar is of particular relevance to geometrization is that in dimension $n = 3$, singularity models corresponding to the soliton metric

$$g_0 = g_\Sigma + dz^2 \tag{4.3}$$

on quotients of $\mathbb{R}^3$ would represent a serious obstacle to proving geometrization via the Ricci flow, because it is not known how to perform surgery at singularities tending to such limits. (We will see the cigar again in Section 5 below.)

Another family of Ricci solitons of special interest comprises the complete metrics $(\mathbb{R}^n, g_0)$ found for $n \geq 3$ by Robert Bryant and Tom Ivey [21]. These are called *gradient solitons* because the vector field X in (4.2) is the gradient field of a potential function. Although the metrics g_0 cannot be written down explicitly, one can compute that the curvature decays like $1/s$ as one moves away from the origin.

In spite of the special properties possessed by singularity models in all dimensions, singularities of the Ricci flow $(\mathcal{M}^n, g(t))$ in high dimensions are expected to be very complex. In dimension $n = 3$ however, there are three observations that lead one to expect singularities to be relatively tractable.

The first observation is a pinching estimate proved independently by Hamilton [16] and by Ivey [20] and later improved by Hamilton [18]. Recall that one may regard the Riemann curvature tensor as a self-adjoint operator

$$\mathrm{Rm} : \wedge^2 T\mathcal{M}^n \to \wedge^2 T\mathcal{M}^n.$$

On a 3-manifold, $\wedge^2 T_x\mathcal{M}^3$ is a three-dimensional vector space for each $x \in \mathcal{M}^3$, so one may denote the eigenvalues of $\mathrm{Rm}(x)$ by $\lambda_1 \leq \lambda_2 \leq \lambda_3$. The *curvature pinching estimate* says that if $(\mathcal{M}^3, g(t))$ is a solution of the Ricci flow for $0 \leq t < T$ such that $\lambda_1 \geq -1$ everywhere on $\mathcal{M}^3$ at $t = 0$ (which can always be achieved by scaling), then at any point $x \in \mathcal{M}^3$ and time $t \in [0, T)$ such that $\lambda_1 < 0$, the scalar curvature satisfies

$$R \geq |\lambda_1| \left(\log |\lambda_1| + \log(1+t) - 3 \right).$$

This estimate says that at any point and time where a sectional curvature is negative and large in absolute value, one finds a much larger positive sectional curvature. It implies in particular that any singularity model $(\mathcal{M}^3_\infty, g_\infty(t), x_\infty)$ must have nonnegative sectional curvature at $t = 0$.

The second observation which restricts the possible singularity models one may see in dimension $n = 3$ is the fact that $\mathfrak{so}(2)$ is the only proper nontrivial Lie subalgebra of $\mathfrak{so}(3)$. This fact says that at the origin of any singularity model, the eigenvalues of the curvature operator (after scaling) must conform to either the signature $(+,+,+)$ or else $(0,0,+)$. The other possible pattern, $(0,0,0,)$, is ruled out by (3.4).

The third observation is the fact that a strong maximum principle holds for tensors. Because the curvature operator of any singularity model $(\mathcal{M}^3_\infty, g_\infty(t), x_\infty)$ is nonnegative at $t = 0$ and has either the signature $(+,+,+)$ or else $(0,0,+)$ at the single point $(x_\infty, 0)$, the strong maximum principle says that the curvature operator must possess either the signature $(+,+,+)$ or else $(0,0,+)$ respectively at *all* points $x \in \mathcal{M}^3_\infty$ and times $t > 0$ such that the limit solution $g_\infty(t)$ exists.

When viewed at an appropriate length scale, the geometry of a solution to the Ricci flow $(\mathcal{M}^3, g(t))$ that becomes singular at time $T < \infty$ closely resembles the singularity model $(\mathcal{M}^3_\infty, g_\infty(t), x_\infty)$ one obtains by blowing-up the singularity, at least for points near the singularity and times just before T.

The standard example of a singularity of signature $(+,+,+)$ is the shrinking round 3-sphere with the metric $r(t)^2 g_{\text{can}}$, where $r(t)$ is given by (3.3) with $n = 3$. As was remarked above, this singularity model itself exhibits a Type I singularity at some finite time.

Singularities of signature $(0,0,+)$ are called *neckpinches*. Near the singularity, a solution which encounters a finite-time neckpinch is expected to resemble the soliton solution

$$g(t) = ds^2 + r(t)^2 g_{\text{can}} \tag{4.4}$$

on $\mathbb{R} \times \mathcal{S}^2$. Here $r(t)$ is given by (3.3) with $n = 2$. Notice that this model also forms a Type I singularity at some time $T < \infty$. Understanding neckpinches is a vital part of current efforts to obtain topological information from the Ricci flow: developing necks should be geometrically and topologically simple enough that one could remove a small piece of a neck just before a singularity forms in such a way that the curvature on the complement of this piece obeys a bound which allows the flow to be continued there, at least for a short time.

Remarkably, there were until recently no rigorous examples of neckpinch singularities on compact manifolds. The first examples of any sort were warped-product metrics constructed by Miles Simon [27] on $\mathbb{R} \times \mathcal{S}^n$. His construction used a supersolution as a barrier to force a singularity to occur in finite time on a compact subset of the manifold. Sigurd Angenent and I have recently studied neckpinch singularities on $\mathcal{S}^{n+1}$ for any $n \geq 2$. We consider metrics of the form

$$g(x,t) = \varphi(x,t)^2\, dx^2 + \psi(x,t)^2\, g_{\text{can}}$$

on $(-1,1) \times \mathcal{S}^n$, which we identify with the sphere $\mathcal{S}^{n+1}$ with its north and south poles removed.

A rough outline of our method is as follows. We call a local minimum of the radius a 'neck' and a local maximum a 'bump'. We consider $\mathrm{SO}(n+1)$-invariant metrics on $\mathcal{S}^{n+1}$ which possess bumps and necks, have positive scalar curvature everywhere, and also obey a gradient bound. One can construct simple examples of such data by removing a neighborhood of the equator of the standard unit sphere and replacing it with a narrow neck. (The figure below shows a reflection-symmetric metric having a single neck and two bumps, but our results are more general.) The hypothesis of $\mathrm{SO}(n+1)$ symmetry implies that the Riemann curvature tensor is completely determined by the sectional curvature K_0 of the 2-planes perpendicular to the spheres $\{x\} \times \mathcal{S}^n$ and the sectional curvature K_1 of the 2-planes tangential to these spheres. We derive pinching estimates for these curvatures which imply that a Type I (rapidly forming) singularity must develop at a neck at some time $T < \infty$. These estimates show that any sequence of parabolic dilations formed at the developing singularity converges to a shrinking cylinder soliton (4.4) on $\mathbb{R} \times \mathcal{S}^n$, with $r(t)$ given by (3.3). We prove that this convergence takes place uniformly in any ball of radius $o\left(\sqrt{-(T-t)\log(T-t)}\right)$ centered at the neck and obtain further estimates for the asymptotics of the developing singularity. (In forthcoming work, we intend to show that these estimates are in fact sharp.)

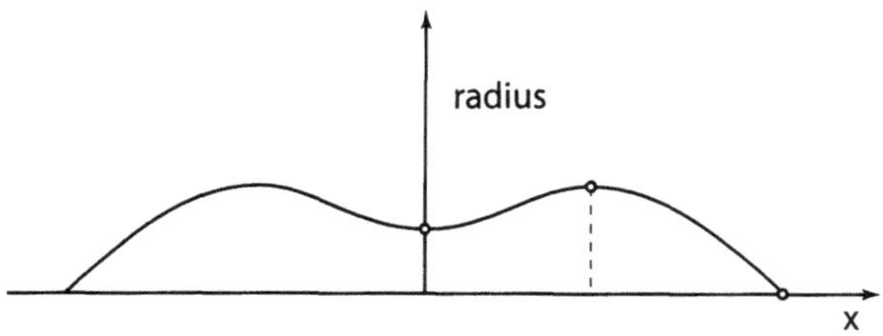

A sphere with one neck and two bumps.

5. The topology of singularities

As was mentioned above, there are reasons to believe that singularity models in dimension $n = 3$ are amenable to classification. In fact, using delicate geometric and analytic arguments, a partial classification was obtained in [16]. Let us now recall the part of that classification that deals with finite-time singularities.

1. If a solution $(\mathcal{M}^3, g(t))$ of the unnormalized Ricci flow encounters a Type I singularity, then after performing dilations correctly and obtaining an injectivity radius estimate, one obtains a limit which is a quotient of either
   (a) a compact shrinking round sphere $(\mathcal{S}^3, g(t))$, where $g(t)$ is given by (3.2), or
   (b) a noncompact shrinking cylinder $(\mathbb{R} \times \mathcal{S}^2, g(t))$, where $g(t)$ is given by (4.4).
2. If a solution $(\mathcal{M}^3, g(t))$ of the unnormalized Ricci flow encounters a Type IIa singularity, then after performing dilations correctly and obtaining an injectivity radius estimate, one gets a quotient of one of the following noncompact limits:
   (a) a translating self-similar solution $(\mathbb{R}^3, g(t))$ where $g(t)$ has the form given in (4.1),
   (b) a shrinking cylinder $(\mathbb{R} \times \mathcal{S}^2, g(t))$ as in Case 1b above, or
   (c) a cigar product $(\mathbb{R}^3, g(t))$, where $g(t)$ is the self-similar solution corresponding to the soliton metric g_0 given in (4.3).

An example of what one might see in Case 2a is the Bryant-Ivey soliton mentioned above. For any limit in Case 2a, one performs what is called *dimension reduction* to obtain an ancient solution. (Dimension reduction is a technique that involves taking a limit around a suitable sequence of points tending to spatial infinity; it will not be discussed further in this introductory survey.) If the ancient solution one obtains is not in fact an eternal solution which attains its maximum curvature, one then takes a third limit about a suitable sequence of points and times tending to $-\infty$ where the curvature is sufficiently near its maximum. Having done so, one sees either Case 2b or 2c above.

Since the limit in Case 1a is compact, the underlying manifold of the original solution must have been $\mathcal{S}^3$ or one of its quotients. In the other cases, the singularity model gives local information about the original solution near the singularity just prior to its formation. The recent work [22] of Perelman rules out Case 2c. This is highly significant, because (as was remarked above) it was not known how to perform surgery on the original solution if this case were to occur.

There are interesting connections between the topology of a manifold and the singularities it admits. Some of these are revealed by the method of performing geometric-topological surgeries just prior to singularity formation. Others may be found by more direct means. To conclude this survey, we offer one example of such a connection. In recent work [19], Tom Ilmanen and I were able to rule out product metrics on $\mathcal{S}^1 \times \mathcal{S}^2$ as possible singularity models, thereby answering affirmatively a conjecture made in [16]. Our result rests on a more general principle which yields a lower bound for the diameter of any solution $(\mathcal{M}^n, g(t))$ to the Ricci flow on certain manifolds of any dimension. Having such a bound implies in particular that any singularity model constructed from a finite-time singularity on one of those manifolds must be noncompact.

Here is an outline of the proof. Given a free homotopy class Γ on a compact Riemannian manifold $(\mathcal{M}^n, g)$, we define

$$\ell_g(\Gamma) = \inf_{\gamma \in \Gamma} \operatorname{length}_g(\gamma)$$

and

$$m_g(\Gamma) = \liminf_{k \to \infty} \frac{\ell_g(k\Gamma)}{k}.$$

Then we prove the monotonicity result that if $(\mathcal{M}^n, g(t))$ is a solution of the Ricci flow, then

$$m_{g(t_1)}(\Gamma) \geq m_{g(t_0)}(\Gamma)$$

for any $t_1 \geq t_0$. In certain cases (for example, if the image of Γ in $H_1(M^n; \mathbb{R})$ is nonzero, or if $\mathcal{M}^3$ is Haken, or if $\pi_1(\mathcal{M}^3)$ is word-hyperbolic) one knows that $m_{g(0)}(\Gamma)$ is positive. (Question: when is $m_{g(0)}(\Gamma)$ positive for a free homotopy class Γ in a homology 3-sphere?) Whenever one knows that $m_{g(0)}(\Gamma)$ is positive, our method yields a lower bound for the diameter of the solution $g(t)$ for however long it exists. In particular, if $\alpha \in H_1(M^n; \mathbb{Z})$ is any element of infinite order, one concludes that the infimum of the lengths measured with respect to $g(t)$ of all curves representing α is bounded below by some $c > 0$ depending only on α and g_0.

This survey has presented only a tiny fraction of what is known by experts in singularity formation for the Ricci flow. As was remarked in the introduction, anyone wishing more detailed information is urged to consult the papers of the original authors.

References

[1] Angenent, Sigurd. Knopf, Dan. *An example of neckpinching for Ricci flow on* S^{n+1}. Preprint.

[2] Bando, Shigetoshi. *Real analyticity of solutions of Hamilton's equation.* Math. Z. **195** (1987), no. 1, 93–97.

[3] Bemelmans, Josef; Min-Oo; Ruh, Ernst A. *Smoothing Riemannian metrics.* Math. Z. **188** (1984), no. 1, 69–74.

[4] Cao, Huai-Dong. *Existence of gradient Kähler-Ricci solitons.* Elliptic and parabolic methods in geometry (Minneapolis, MN, 1994), 1–16, A K Peters, Wellesley, MA, 1996.

[5] Cao, Huai-Dong. *Limits of solutions to the Kähler-Ricci flow.* J. Differential Geom. **45** (1997), no. 2, 257–272.

[6] Cao, Huai-Dong and Chow, Bennett. *Recent developments on the Ricci flow.* Bull. Amer. Math. Soc. (N.S.) **36** (1999), no. 1, 59–74.

[7] Chow, Bennett. *A Survey of Hamilton's Program for the Ricci Flow on 3-manifolds.* `arXiv:math.DG/0211266`.

[8] Chow, Bennett; Knopf, Dan; Lu, Peng. *Hamilton's injectivity radius estimate for sequences with almost nonnegative curvature operators.* Comm. Anal. Geom. **10** (2002), no. 5, 1151–1180.

[9] Chu, Sun-Chin. *Geometry of 3-Dimensional Gradient Ricci Solitons with Positive Curvature.* `arXiv:math.DG/0303135`.

[10] DeTurck, Dennis M. *Deforming metrics in the direction of their Ricci tensors.* J. Differential Geom. **18** (1983), no. 1, 157–162.

[11] DeTurck, Dennis M. *Deforming metrics in the direction of their Ricci tensors, improved version,* Selected Papers on the Ricci Flow, ed. H.-D. Cao, B. Chow, S.-C. Chu, and S.-T. Yau, International Press. To appear.

[12] Feldman, Mikhail; Ilmanen, Tom; Knopf, Dan. *Rotationally symmetric shrinking and expanding gradient Kähler-Ricci solitons.* Preprint.

[13] Guenther, Christine; Isenberg, James; Knopf, Dan. *Stability of the Ricci flow at Ricci-flat metrics.* Comm. Anal. Geom. **10** (2002), no. 4, 741–777.

[14] Hamilton, Richard S. *Three-manifolds with positive Ricci curvature.* J. Differential Geom. **17** (1982), no. 2, 255–306.

[15] Hamilton, Richard S. *Four-manifolds with positive curvature operator.* J. Differential Geom. **24** (1986), no. 2, 153–179.

[16] Hamilton, Richard S. *The formation of singularities in the Ricci flow.* Surveys in differential geometry, Vol. II (Cambridge, MA, 1993), 7–136, Internat. Press, Cambridge, MA, 1995.

[17] Hamilton, Richard S. *A compactness property for solutions of the Ricci flow.* Amer. J. Math. **117** (1995), no. 3, 545–572.

[18] Hamilton, Richard S. *Non-singular solutions of the Ricci flow on three-manifolds.* Comm. Anal. Geom. **7** (1999), no. 4, 695–729.

[19] Ilmanen, Tom; Knopf, Dan. *A lower bound for the diameter of solutions to the Ricci flow with nonzero* $H^1(M^n;\mathbb{R})$. Math. Res. Lett. **10** (2003) 161–168.

[20] Ivey, Thomas. *Ricci solitons on compact three-manifolds.* Differential Geom. Appl. **3** (1993), no. 4, 301–307.

[21] Ivey, Thomas. *New examples of complete Ricci solitons.* Proc. Amer. Math. Soc. **122** (1994), no. 1, 241–245.

[22] Perelman, Grisha. *The entropy formula for the Ricci flow and its geometric applications.* `arXiv:math.DG/0211159`.

[23] Perelman, Grisha. *Ricci flow with surgery on three-manifolds.* `arXiv:math.DG/0303109`.

[24] Scott, Peter. *The geometries of 3-manifolds.* Bull. London Math. Soc. **15** (1983), no. 5, 401–487.

[25] Shi, Wan-Xiong. *Deforming the metric on complete Riemannian manifolds.* J. Differential Geom. **30** (1989), no. 1, 223–301.

[26] Shi, Wan-Xiong. *Ricci deformation of the metric on complete noncompact Riemannian manifolds.* J. Differential Geom. **30** (1989), no. 2, 303–394.

[27] Simon, Miles. *A class of Riemannian manifolds that pinch when evolved by Ricci flow.* Manuscripta Math. **101** (2000), no. 1, 89–114.

[28] Thurston, William P. *Three-dimensional manifolds, Kleinian groups and hyperbolic geometry.* Bull. Amer. Math. Soc. (N.S.) **6** (1982), no. 3, 357–381.

Dan Knopf
The University of Iowa
e-mail: `dknopf@math.uiowa.edu`
URL: `http://www.math.uiowa.edu/~dknopf`

Progress in Nonlinear Differential Equations
and Their Applications, Vol. 59, 81–93

A Family of Expanding Ricci Solitons

Andreas Gastel and Manfred Kronz

0. Introduction

The Ricci flow is a natural evolution equation for Riemannian metrics, introduced by Richard S. Hamilton in 1982. A family $(g(t,\cdot))_{t\in I}$ of metrics on a Riemannian manifold M, depending on a time parameter $t \in I \subseteq \mathbb{R}$ is a solution to the Ricci flow if it solves the equation

$$\frac{\partial}{\partial t} g(t,\cdot) = -2\mathrm{Ric}\, g(t,\cdot),$$

where $\mathrm{Ric}\, g(t,\cdot)$ is the Ricci tensor associated with the evolving metric $g(t,\cdot)$. In general, a solution of the Ricci flow starting with smooth initial data will not possess a smooth continuation for all time. The formation of singularities has been discussed extensively in Hamilton's article [H], and there is a particular type of solutions which is expected (and in some cases known) to appear as parabolic blow-up limit of Ricci flows around a singularity, namely *Ricci solitons.* Ricci solitons are solutions to the Ricci flow for which there exist scalars $\sigma(t)$ and diffeomorphisms $\Psi_t : M \to M$ such that

$$g(t,\cdot) = \sigma(t)\, \Psi_t^* g(T,\cdot) \quad \text{for all } t \in I \text{ and some fixed } T \in I. \tag{1}$$

This holds if and only if there is a vector field X on M and a constant $\lambda \in \mathbb{R}$ such that

$$-2\mathrm{Ric}\, g(T,\cdot) = \mathcal{L}_X g(T,\cdot) + \lambda\, g(T,\cdot)\,, \tag{2}$$

where $\mathcal{L}_X$ is the Lie derivative in the direction of X; see [FIK, Section 2.1] for a proof of the equivalence of (1) and (2) and for a list of references of examples of Ricci solitons. The cases $\lambda < 0$, $\lambda = 0$, and $\lambda > 0$ correspond to solitons *of shrinking, translating*, and *expanding type*, respectively. In this note, we restrict ourselves to doubly warped product metrics of the form

$$g(t,x,y,z) = h(t,x)^2 \oplus r(t,x)^2 \gamma_m(y) \oplus \rho(t,x)^2 \gamma_n(z), \tag{3}$$

on $M := \mathbb{R}_+ \times S^m \times N$, where (S^m, γ_m), $m \geq 2$, is the m-dimensional sphere with the canonical metric, and (N, γ_n) is an Einstein manifold of dimension $n \geq 2$ having Einstein constant $\kappa > 0$. In addition to (3), we require the solutions to close smoothly at the lower end of $\mathbb{R}_+$ to give *complete Ricci solitons on* $\mathbb{R}^{m+1} \times N$. Our motivation is the fact that there are rather similar solitons for the mean curvature flow which lead to nonuniqueness; see the examples in [AIV]. (See also

[G] for an analogous example for the Yang-Mills heat flow.) Translating solitons of the form described here have been constructed by Ivey [Iv], and Einstein metrics (i.e., stationary solutions) by Böhm [B2]. In this note we construct a *two-parameter family of expanding solitons.* They exist for $t > 0$ and develop from a cone at $t = 0$. Since they depend essentially on only one variable, they are actually expanding *gradient* Ricci solitons, which means X in (2) is the gradient vector field of some function on M. See the next section and Theorem 1 in Section 5 for more details. We think that there should be also ***shrinking*** solitons with the same symmetries, but they seem to be harder to find. We wish to thank Frank Duzaar and Joseph F. Grotowski for interesting discussions.

1. The reduced equation

We consider doubly warped product metrics of the form (3). Using the fact that both S^m and N are Einstein, we can calculate the Ricci curvature in terms of h, r and ρ to reduce the Ricci flow equation $\frac{\partial}{\partial t}g(t,\cdot) = -2\mathrm{Ric}\, g(t,\cdot)$ to the following system denoted by **(R1)**:

$$\begin{aligned} h_t &= \frac{m}{r}\Big(\frac{r_{xx}}{h} - \frac{h_x r_x}{h^2}\Big) + \frac{n}{\rho}\Big(\frac{\rho_{xx}}{h} - \frac{h_x\rho_x}{h^2}\Big), \\ r_t &= \frac{r_{xx}}{h^2} - \frac{h_x r_x}{h^3} + (m-1)\frac{r_x^2}{h^2 r} - \frac{m-1}{r} + n\,\frac{\rho_x r_x}{h^2\rho}, \\ \rho_t &= \frac{\rho_{xx}}{h^2} - \frac{h_x\rho_x}{h^3} + (n-1)\frac{\rho_x^2}{h^2\rho} - \frac{\kappa}{\rho} + m\,\frac{r_x\rho_x}{h^2 r}. \end{aligned}$$

For further simplification, we reparametrize the first factor by arc length, $\xi(t,x) := \int_0^x h(t,y)\,dy$, and restrict to a *parabolically self-similar* ansatz. This means that we assume h, r, ρ to be of the form

$$h(t,x) = \chi\Big(\frac{\xi(t,x)}{\sqrt{\mu t}}\Big), \qquad r(t,x) = \sqrt{\mu t}\,\varphi\Big(\frac{\xi(t,x)}{\sqrt{\mu t}}\Big), \qquad \rho(t,x) = \sqrt{\mu t}\,\psi\Big(\frac{\xi(t,x)}{\sqrt{\mu t}}\Big)$$

for some functions φ, ψ, χ of one variable, with $\mu = \pm 1$. If $\mu = 1$, such solutions are defined for $t > 0$ and can be viewed as "self-expanding", while in the case $\mu = -1$ they are defined for $t < 0$ and are "self-shrinking". Introducing the new variable $s := \frac{\xi(t,x)}{\sqrt{\mu t}}$, we find

$$h_x = \frac{\chi(s)\,\chi'(s)}{\sqrt{\mu t}}, \qquad r_x = \chi(s)\varphi'(s), \qquad \rho_x = \chi(s)\psi'(s),$$

$$\frac{r_{xx}}{h} - \frac{h_x r_x}{h^2} = \frac{\chi(s)\,\varphi''(s)}{\sqrt{\mu t}}, \qquad \frac{\rho_{xx}}{h} - \frac{h_x\rho_x}{h^2} = \frac{\chi(s)\,\psi''(s)}{\sqrt{\mu t}}.$$

To calculate the time derivatives, we first find that $\frac{\partial\xi(t,x)}{\partial t}$ is, by the h_t-equation,

$$\frac{\partial\xi}{\partial t} = \int_0^x h_t(t,y)\,dy = \frac{m}{\sqrt{\mu t}}\int_0^s \frac{\varphi''(\sigma)}{\varphi(\sigma)}\,d\sigma + \frac{n}{\sqrt{\mu t}}\int_0^s \frac{\psi''(\sigma)}{\psi(\sigma)}\,d\sigma =: \frac{1}{\sqrt{\mu t}}\,I(\varphi,\psi;s).$$

Hence we have $h_t(t,x) = \frac{1}{\mu t}(I(\varphi,\psi;s) - \frac{\mu}{2}s)\chi'(s)$ and

$$\begin{aligned} r_t(t,x) &= \frac{1}{\sqrt{\mu t}}\Big(-\frac{\mu}{2}s\varphi'(s) + I(\varphi,\psi;s)\varphi'(s) + \frac{\mu}{2}\varphi(s)\Big), \\ \rho_t(t,x) &= \frac{1}{\sqrt{\mu t}}\Big(-\frac{\mu}{2}s\psi'(s) + I(\varphi,\psi;s)\psi'(s) + \frac{\mu}{2}\psi(s)\Big). \end{aligned}$$

Inserting this into (R1) we can eliminate χ from the system, i.e.,

$$\chi(s) = \chi(0)\exp\Big(\int_0^s \frac{m\frac{\varphi''}{\varphi} + n\frac{\psi''}{\psi}}{I(\varphi,\psi;\sigma) - \frac{\mu}{2}\sigma}\,d\sigma\Big),$$

which leaves us with the system **(R2)**:

$$\begin{aligned} &\Big(\frac{\varphi'}{\varphi}\Big)' + \Big(m\frac{\varphi'}{\varphi} + n\frac{\psi'}{\psi} + \frac{\mu}{2}s - I(\varphi,\psi)\Big)\frac{\varphi'}{\varphi} - \frac{\mu}{2} - \frac{m-1}{\varphi^2} = 0, \\ &\Big(\frac{\psi'}{\psi}\Big)' + \Big(m\frac{\varphi'}{\varphi} + n\frac{\psi'}{\psi} + \frac{\mu}{2}s - I(\varphi,\psi)\Big)\frac{\psi'}{\psi} - \frac{\mu}{2} - \frac{\kappa}{\psi^2} = 0. \end{aligned}$$

We can always assume $\kappa = n-1$ (which is $N = S^n$) after replacing ψ by $\sqrt{\frac{\kappa}{n-1}}\psi$. If $\mu = 1$, the Riemannian metrics $g(t,x,y,z) = \chi(\frac{\xi(t,x)}{\sqrt{\mu t}})^2 \oplus \mu t\varphi(\frac{\xi(t,x)}{\sqrt{\mu t}})^2\gamma_m(y) \oplus \mu t\psi(\frac{\xi(t,x)}{\sqrt{\mu t}})^2\gamma_n(z)$ describe an expanding Ricci soliton, $g(t,\cdot) = \sigma(t)\Psi_t^* g(1,\cdot)$, where

$$\sigma(t) = t, \qquad \Psi_t(x,y,z) = \Big(\xi(1,\cdot)^{-1}\Big(\frac{\xi(t,x)}{\sqrt{\mu t}}\Big), y, z\Big).$$

Using the second Bianchi identity $2g^{ij}D_jR_{ik} = D_kR$, we can integrate I explicitly. For a constant $c \in \mathbb{R}$, we have

$$I(\varphi,\psi) = m(m-1)\frac{\varphi'^2 - 1}{2\varphi^2} + n(n-1)\frac{\psi'^2 - 1}{2\psi^2} + mn\frac{\varphi'\psi'}{\varphi\psi} + c.$$

For all $\mu \in \mathbb{R}$, there is an explicit solution to the system (R2),

$$\varphi_{\text{cone}}(s) = \sqrt{\frac{m-1}{m+n-1}}\,s, \quad \psi_{\text{cone}}(s) = \sqrt{\frac{n-1}{m+n-1}}\,s,$$

which we call the *Ricci-flat cone*. We are interested in solutions close to this cone, which motivates the transformation

$$\begin{aligned} p(\log s) &:= m\log\Big(\frac{\varphi(s)}{\sqrt{\frac{m-1}{m+n-1}}\,s}\Big) + n\log\Big(\frac{\psi(s)}{\sqrt{\frac{n-1}{m+n-1}}\,s}\Big), \\ q(\log s) &:= \log\Big(\sqrt{\frac{n-1}{m-1}}\frac{\varphi(s)}{\psi(s)}\Big). \end{aligned}$$

Now we have the system **(R3)**:

$$
\begin{aligned}
J(p,q;y) &:= e^y \int_{-\infty}^{y} \Big(p'' + p' + \frac{p'^2 + mnq'^2}{m+n}\Big) e^{-x}\,dx,\\
p'' &+ \Big(2m + 2n - 1 + \frac{\mu}{2} e^{2y} - J(p,q) + p'\Big)p' - (m+n)J(p,q)\\
&+ (m+n-1)\Big[m\Big(1 - e^{-\frac{2(p+nq)}{m+n}}\Big) + n\Big(1 - e^{-\frac{2(p-mq)}{m+n}}\Big)\Big] = 0,\\
q'' &+ \Big(m + n - 1 + \frac{\mu}{2} e^{2y} - J(p,q) + p'\Big)q'\\
&+ (m+n-1)\Big[e^{-\frac{2(p-mq)}{m+n}} - e^{-\frac{2(p+nq)}{m+n}}\Big] = 0,
\end{aligned}
$$

and the Ricci-flat cone is simply $p_{\text{cone}} \equiv 0$, $q_{\text{cone}} \equiv 0$. There is a natural energy quantity associated with (R3), namely

$$
W := \frac{m+n-1}{2}\Big(n(e^{-\frac{2(p-mq)}{m+n}} - 1) + m(e^{-\frac{2(p+nq)}{m+n}} - 1) + 2p\Big) + \frac{p'^2 + mnq'^2}{2(m+n)}. \tag{4}
$$

We have

$$
W' = (J - p')p' - \Big(m + n - 1 + \frac{\mu}{2} e^{2y} - (J - p')\Big)\frac{p'^2 + mnq'^2}{m+n}. \tag{5}
$$

We are interested in this quantity because W' is fairly simple and $W \to 0$ is equivalent to $p \to 0$, $q \to 0$, $p' \to 0$, $q' \to 0$. For the quantity $(J - p')$, which appears twice in the expression for W' we have, by the definition of J, the relation

$$
(J - p')' = (J - p') + 2p' + \frac{p'^2 + mnq'^2}{m+n}. \tag{6}
$$

2. The initial value problem

Lemma 1. *For each $\mu \in \mathbb{R}$, each $a \in \mathbb{R}$ and each $b > 0$, there exists a unique solution $(\varphi_{\mu,a,b}, \psi_{\mu,a,b})$ of (R2) with initial values*

$$
\begin{aligned}
&\varphi_{\mu,a,b}(0) = 0, \quad \varphi'_{\mu,a,b}(0) = 1, \quad \varphi''_{\mu,a,b}(0) = 0, \quad \varphi'''_{\mu,a,b}(0) = 6a,\\
&\psi_{\mu,a,b}(0) = b, \quad \psi'_{\mu,a,b}(0) = 0.
\end{aligned}
$$

The functions $\varphi_{\mu,a,b}, \psi_{\mu,a,b}$ and their derivatives depend Lipschitz-continuously on a, b, and μ. Furthermore, $\varphi_{\mu,a,b}$ is odd and $\psi_{\mu,a,b}$ is even.

Remark. The functions $\varphi_{\mu,a,b}, \psi_{\mu,a,b}$ correspond to a solution of the Ricci flow on $\mathbb{R}^{m+1} \times N$. The initial values and the parities from the lemma imply that the warped product $\mathbb{R}_+ \times S^m \times N$ closes smoothly at the lower end of $\mathbb{R}_+$.

Proof of the Lemma. Here we use the following proposition first proved by [EW], and formulated in a convenient way in [B1], [B2]. We need a slightly more general version, which is also proved in [EW], except for the Lipschitz-continuous dependence, which can be proved as in [FK]. The proof of the proposition is based on formal power series expansions.

Proposition 1 ([EW]). *Consider the differential equation*

$$y'(s) = A(s, y(s)) + \frac{1}{s}\, B(y(s)),$$

where $A : \mathbb{R}_+ \times U \to \mathbb{R}^\ell$, $B : U \to \mathbb{R}^\ell$ *are analytic functions on an open neighborhood* U *of* $w \in \mathbb{R}^\ell$. *If* $B(w) = 0$ *and* $I_\ell - \frac{1}{j} DB(w)$ *is invertible for all* $j \in \mathbb{N}$, *there is a unique local solution* $y(s)$ *of the differential equation with* $y(0) = w$. *Moreover,* $y(s)$ *depends Lipschitz-continuously on* $w \in \{v \in U | B(v) = 0\}$. □

First we have to transform (R2) to make Proposition 1 applicable. We let $y_1 := s^{-1}\varphi$, $y_2 := s^{-2}(\varphi' - 1)$, $y_3 := \psi$, $y_4 := s^{-1}\psi'$, $y_5 := s^{-1} I(\varphi, \psi)$ in order to obtain $y'(s) = A(s, y(s)) + \frac{1}{s}\, B(y(s))$ for some A as above, where

$$B(y) = \begin{pmatrix} 1 - y_1 \\ 2(1-m)\frac{y_2}{y_1} - n\frac{y_4}{y_3} - \frac{\mu}{2} + y_5 + \frac{\mu y_1}{2} - 2y_2 \\ 0 \\ (n-1)\frac{1}{y_3} - m\frac{y_4}{y_1} + \frac{\mu y_3}{2} - y_4 \\ 2m(1-m)\frac{y_2}{y_1^2} + n(n-1)\frac{1}{y_3^2} - 2mn\frac{y_4}{y_1 y_3} - m\frac{\mu}{2y_1} + m\frac{y_5}{y_1} + \frac{\mu(m+n)}{2} - y_5 \end{pmatrix}.$$

All zeroes of B are of the form

$$w = \begin{pmatrix} 1 \\ 3a \\ b \\ \frac{1}{m+1}\left(\frac{n-1}{b} + \frac{\mu}{2}\, b\right) \\ 6ma + \frac{n}{m+1}\left(\frac{n-1}{b^2} + \frac{\mu}{2}\right) \end{pmatrix}$$

where a and b are as in the lemma. Up to entries which we do not need and denote by $*$, we calculate

$$I_5 - \tfrac{1}{j} DB(w) = \begin{pmatrix} 1 + \frac{1}{j} & 0 & 0 & 0 & 0 \\ * & 1 + \frac{2m}{j} & * & * & -\frac{1}{j} \\ 0 & 0 & 1 & 0 & 0 \\ * & 0 & * & 1 + \frac{m+1}{j} & 0 \\ * & \frac{2m(m-1)}{j} & * & * & 1 - \frac{m-1}{j} \end{pmatrix}$$

for every zero w of B. Hence

$$\det(I_5 - \tfrac{1}{j} DB(w)) = (1 + \tfrac{1}{j})(1 + \tfrac{m+1}{j})^2 \neq 0$$

for all $j \in \mathbb{N}$, and Proposition 1 can be applied. This means there is a 2-parameter family of local solutions depending Lipschitz-continuously on a, b. It is more standard to show that the μ-dependence is also Lipschitz. From (R2) it is easy to read off that $\varphi_{\mu,a,b}$ is odd and $\psi_{\mu,a,b}$ is even, which completes the proof of Lemma 1. □

Transforming from (R2) to (R3), we denote the corresponding solutions by $p_{\mu,a,b}$, $q_{\mu,a,b}$.

Corollary 1. *For $\mu, a \in \mathbb{R}$, $b > 0$, and $\tau > 0$, we have*

$$(p_{\mu,a,b}, q_{\mu,a,b})(y) = (p_{\tau^2\mu,\tau^2 a,\tau^{-1}b}, q_{\tau^2\mu,\tau^2 a,\tau^{-1}b})(y - \log \tau).$$

Proof. We check that $(\varphi_{\mu,a,b}, \psi_{\mu,a,b})(s)$ and $(\tau\varphi_{\tau^2\mu,\tau^2 a,\tau^{-1}b}, \tau\psi_{\tau^2\mu,\tau^2 a,\tau^{-1}b})(\tau^{-1}s)$ solve the same initial value problem in Lemma 1, hence are equal by uniqueness. Transformation to (R3) yields the assertion. □

3. A Ricci-flat manifold resolving the cone singularity

A time-independent solution to the Ricci flow is Ricci flat. If this solution is of the form (3) on $\mathbb{R}_+ \times S^m \times N$, we can replace (r, ρ, x) by (φ, ψ, s) and find that it solves (R2) with $\mu = 0$, and further there holds $I(\varphi, \psi) \equiv 0$. It was proved in [EW] that this additional equation follows from the other two if suitable initial values are chosen. Transforming as above, we obtain (R3) with $\mu = 0$ and $J \equiv 0$. Up to rescaling, there is only one nontrivial solution corresponding to the initial value problem from Lemma 1. Fixing $b = 1$, we find that $I \equiv 0$ implies $a = -\frac{n(n-1)}{6m(m+1)}$. We denote the corresponding solution of (R3) by (p_0, q_0). The following lemma proves its long-time existence, which also follows from the more general existence theorem in [B2].

Lemma 2. *Let $m, n \geq 2$. The solution (p_0, q_0) of (R3) exists on $\mathbb{R}$ and has asymptotics*

$$p_0'(y) = -n + O(e^y), \qquad q_0'(y) = 1 + O(e^y),$$
$$e^{-\frac{2(p_0 - mq_0)}{m+n}} = \frac{m-1}{m+n-1} + O(e^y), \qquad e^{-\frac{2(p_0+nq_0)}{m+n}} = O(e^y)$$

for $y \to -\infty$, while for $y \to \infty$ they are

$$p_0(y)^2 + p_0'(y)^2 = O(e^{-2y}), \qquad q_0(y)^2 + q_0'(y)^2 = O(e^{-3y/2}),$$
$$e^{-\frac{2(p_0 - mq_0)}{m+n}} = 1 + O(e^{-3y/4}), \qquad e^{-\frac{2(p_0+nq_0)}{m+n}} = 1 + O(e^{-3y/4}).$$

Proof. Local existence has been proved in Lemma 1. The asymptotics at $y \to -\infty$ follow from the initial values for (φ_0, ψ_0). At every y_0 with $p_0'(y) = -(m+n-1)$, the second equation of (R3) implies $p_0''(y_0) > 0$. Since $p'(-\infty) = -n$, it follows that

$$p_0' > -(m+n-1) \text{ on the domain of } (p_0, q_0). \tag{7}$$

Calculating W from (4) for (p_0, q_0), we find

$$W' = -p_0'^2 - (m+n-1+p_0')\frac{p_0'^2 + mnq_0'^2}{m+n}, \tag{8}$$

and hence $W' \leq 0$ on the domain of (p_0, q_0) by (7). Since $W \to \infty$ at some finite time is the only possibility for the domain of (p_0, q_0) to be bounded, p_0, q_0, p_0', q_0' must exist and be bounded. By monotonicity of W, we have $W'(\infty) = 0$, and hence there holds $p_0'(\infty) = q_0'(\infty) = 0$. Plugging this into (R3), we find $p_0''(\infty) = 0$, which in turn implies that the term involving p_0 and q_0 in its second equation goes to 0 as $y \to \infty$. Since this term is also in W, we find existence of $p_0(\infty)$ and $q_0(\infty)$, both of which can only be 0 again by (R3). Now that we know that p_0, q_0, p_0', q_0' are small for $y \to \infty$, the asymptotics follow by linearization of (R3). □

4. Asymptotics

Proposition 2. *Let $m, n \geq 2$. For every fixed $a \in \mathbb{R}$, we have*

$$\lim_{b \searrow 0} (p_{1,a,b}, q_{1,a,b}) \equiv (p_{\text{cone}}, q_{\text{cone}}) \equiv (0,0) \quad \textit{pointwise in } C^1.$$

Proof. From Corollary 1, applied with $\mu = 1$ and $\tau = b$, we infer that

$$(p_{1,a,b}, q_{1,a,b})(y) = (p_{b^2,b^2a,1}, q_{b^2,b^2a,1})(y - \log b).$$

With p_0, q_0 from Lemma 2, we define

$$P_b := p_{b^2,b^2a,1} - p_0, \quad Q_b := q_{b^2,b^2a,1} - q_0, \quad J_b := J(p_{b^2,b^2a,1}, q_{b^2,b^2a,1}).$$

From (R3) we conclude that $w_b := (J_b, P_b, P_b', Q_b, Q_b')$ is a solution of

$$w_b'(y) = \Big(\mathfrak{M}(y) - \frac{b^2}{2}\, e^{2y}\mathfrak{N}\Big) w_b(y) + O\Big(\frac{b^2 e^{2y}}{1+e^y}\Big) + HOT, \tag{9}$$

where $\mathfrak{M}(y)$ stands for

$$\begin{pmatrix} m{+}n{+}1{+}p_0' & -\frac{2d}{m+n}(mZ_n{+}nZ_m) & -2d(1+\frac{p_0'}{m+n}) & \frac{2mnd}{m+n}(Z_m{-}Z_n) & \frac{2mn}{m+n}q_0' \\ 0 & 0 & 1 & 0 & 0 \\ m{+}n{+}p_0' & -\frac{2d}{m+n}(mZ_n{+}nZ_m) & -(2d{+}1{+}2p_0') & \frac{2mnd}{m+n}(Z_m{-}Z_n) & 0 \\ 0 & 0 & 0 & 0 & 1 \\ q_0' & \frac{2d}{m+n}(Z_m{-}Z_n) & -q_0' & -\frac{2d}{m+n}(mZ_m{+}nZ_n) & -(d{+}p_0') \end{pmatrix}$$

and $\mathfrak{N}$ is the matrix

$$\mathfrak{N} := \begin{pmatrix} 0 & 0 & 1 & 0 & 0 \\ 0 & 0 & 0 & 0 & 0 \\ 0 & 0 & 1 & 0 & 0 \\ 0 & 0 & 0 & 0 & 0 \\ 0 & 0 & 0 & 0 & 1 \end{pmatrix}.$$

Here, in the definition of $\mathfrak{M}(y)$ we have abbreviated:

$$d := m{+}n{-}1, \quad Z_m := e^{-\frac{2(p_0 - mq_0)}{m+n}}, \quad Z_n := e^{-\frac{2(p_0+nq_0)}{m+n}}.$$

We first discuss the behaviour of $\mathfrak{M}$ at $-\infty$ and ∞. In view of Lemma 2 we have

$$\mathfrak{M}(-\infty) = \begin{pmatrix} m{+}1 & -\frac{2m(m-1)}{m+n} & -\frac{2m(m+n-1)}{m+n} & -\frac{2mn(m-1)}{m+n} & \frac{2mn}{m+n} \\ 0 & 0 & 1 & 0 & 0 \\ m & -\frac{2(m-1)}{m+n} & -(2m{-}1) & -\frac{2mn(m-1)}{m+n} & 0 \\ 0 & 0 & 0 & 0 & 1 \\ 1 & -\frac{2(m-1)}{m+n} & -1 & -\frac{2n(m-1)}{m+n} & -(m{-}1) \end{pmatrix}.$$

The eigenvalues of this matrix are given by

$$\operatorname{spec}\mathfrak{M}(-\infty) = \{-(m-1), -1, 0, 2\},$$

where the eigenvalue $-(m-1)$ has multiplicity 2. In particular we have

$$\max\operatorname{spec}\mathfrak{M}(-\infty) = 2. \tag{10}$$

A further application of Lemma 2 shows

$$\mathfrak{M}(\infty) = \begin{pmatrix} m+n+1 & -2(m+n-1) & -2(m+n-1) & 0 & 0 \\ 0 & 0 & 1 & 0 & 0 \\ m+n & -2(m+n-1) & -2m-2n+1 & 0 & 0 \\ 0 & 0 & 0 & 0 & 1 \\ 0 & 0 & 0 & -2(m+n-1) & -(m+n-1) \end{pmatrix},$$

where now the eigenvalues of this matrix are given by

$$\operatorname{spec}\mathfrak{M}(\infty) = \{1-m-n, \tfrac{1-m-n}{2} \pm \sqrt{\tfrac{(m+n-9)(m+n-1)}{2}}, -1, 2\}.$$

Therefore we infer

$$\max\operatorname{spec}\mathfrak{M}(\infty) = 2. \tag{11}$$

In view of (10), (11) and Lemma 2 we conclude that there exists a constant c_1 depending only on m and n such that there holds

$$|||\mathfrak{M}(y)||| := \max\operatorname{spec}\mathfrak{M}(y) \le 2 + \frac{c_1}{e^{-y/5} + e^{3y/20}} \tag{12}$$

for any $y \in \mathbb{R}$. From the differential equation (9) we derive the differential inequality

$$|w_b(y)|' \le |||\mathfrak{M}(y) + \frac{b^2}{2}\, e^{2y}\mathfrak{N}|||\; |w_b(y)| + c_2\,\frac{b^2 e^{2y}}{1+e^y} + c_3|w_b(y)|^2 \tag{13}$$

with constants c_2 and c_3 depending only on m and n. Fixing $y_0 \in \mathbb{R}$ and abbreviating

$$A_b(y) := \int_{y_0}^{y} |||\mathfrak{M}(z) + \frac{b^2}{2}\, e^{2z}\mathfrak{N}|||\, dz,$$

we conclude from (13) and Gronwall's inequality that there holds

$$|w_b(y)| \le e^{A_b(y)}|w_b(y_0)| + \int_{y_0}^{y} e^{A_b(y)-A_b(z)}\Big(c_2\,\frac{b^2 e^{2z}}{1+e^z} + c_3|w_b(z)|^2\Big)\, dz. \tag{14}$$

Now, in view of (12), there exist a constant $c_4 = c_4(m,n)$ independent of y_0 and b such that

$$A_b(y) - A_b(z) \le 2(y-z) + c_4(1 + b^2 e^{2y}) \tag{15}$$

for any $y \ge z \ge y_0$. By Lemma 1, i.e., the fact that $\varphi_{b^2,b^2a,1}$, $\psi_{b^2,b^2a,1}$ and $J(\varphi_{b^2,b^2a,1})$ together with their derivatives depend Lipschitz-continuously on the parameters b^2 and $b^2\,a$, we infer that for any $y_0 \le 0$ there exist a constant $c_5(y_0)$ such that $|w_b(y_0)| \le c_5(y_0)b^2$ for any sufficiently small $0 < b \ll 1$. Further, for y sufficiently close to $-\infty$, w_b behaves asymptotically like a solution of the linear system $w_b' = \mathfrak{M}(-\infty)\, w_b$ which implies $|w_b| \sim \frac{b^2}{2}e^{2y}$. Therefore we conclude that

there exists a constant c_6 depending only on m and n such that $c_5(y_0) \le c_6 e^{2y_0}$. Consequently we have

$$|w_b(y_0)| \le c_6 b^2 e^{2y_0} \tag{16}$$

for any $y_0 \le 0$ and any sufficiently small $0 < b \ll 1$.

Claim. *There exist $y_1 \le -1$ and constants $c_7 \ge 1$, $b_0 \in (0,1)$ depending only on m und n such that for any $b \in (0, b_0]$, any $y_0 \le y_1$ and any $y \in [y_0, y_0 + |\log b|]$ there holds*

$$|w_b(y)| \le c_7 |y_0| b^2 e^{2y}.$$

Proof of the claim. We will show that as long as

$$|w_b(y)| \le 2c_7 |y_0| b^2 e^{2y} \tag{17}$$

we in fact have $|w_b(y)| \le c_7|y_0|b^2e^{2y}$ for y_0, y and b given in the claim. Using the assumption (17) (on $[y_0, y]$) we derive via (14), (15), (16) and the fact that $|y_0| \ge 1$

$$|w_b(y)| \le e^{2c_4}\Big(c_6 + 2c_2 + 4c_3c_7^2|y_0|e^{2y_0}\Big)|y_0|\, b^2 e^{2y}. \tag{18}$$

We now choose $y_1 \le -1$ sufficiently small such that

$$2e^{2c_4}(c_6 + 2c_2) \le \frac{1}{8c_3|y_1|}\, e^{-2c_4} e^{-2y_1},$$

and fix

$$c_7 \in \Big[2e^{2c_4}(c_6 + 2c_2), \frac{1}{8c_3|y_1|}\, e^{-2c_4} e^{-2y_1}\Big]$$

(in particular we have $c_7 \ge c_6$). With these choices of y_1 and c the right-hand side (18) is bounded by $c_7|y_0|b^2e^{2y}$ for any $y_0 \le y_1 \le -1$, and this implies the statement of the claim, provided $|w_b(y_0)| \le 2c_7|y_0|^2 b^2 e^{2y_0}$. But this estimate holds because of (16), $|y_0| \ge 1$ and $c_6 \le c_7$. □

From this claim together with Lemma 2 and Corollary 1 we deduce

$$|p_{1,a,b}(y_0)| \le |p_0(y_0 + |\log b|)| + |P_b(y_0 + |\log b|)| \le c_8 b + c_7 |y_0| b^2 e^{2y_0} \tag{19}$$

for any $y_0 \le y_1$ and $b \in (0, b_0]$. Thus $p_{1,a,b}$ is uniformly bounded on each compact interval $I \subset (-\infty, y_1]$, and the same assertion holds for $p'_{1,a,b}$, $q_{1,a,b}$ und $q'_{1,a,b}$. Therefore, as $b_i \searrow 0$, $(p_{1,a,b_i}, q_{1,a,b_i})$ converges to a solution $(\tilde{p}, \tilde{q})$ of (R3) (with $\mu = 1$). In view of (19) and its analogues for p', q, q', J we have

$$\tilde{p}(-\infty) = \tilde{p}'(-\infty) = \tilde{q}(-\infty) = \tilde{q}'(-\infty) = 0.$$

Now, arguing along the lines of the proof of Lemma 1, we see that the only solution of the initial value problem $\varphi(0) = 0$, $\psi(0) = 0$, $\varphi'(0) = \sqrt{\frac{m-1}{m+n-1}}$, $\psi'(0) = \sqrt{\frac{n-1}{m+n-1}}$ is the Ricci-flat cone $(\phi, \psi) \equiv (\varphi_{\text{cone}}, \psi_{\text{cone}})$. This initial data for φ and ψ corresponds to the above-mentioned initial values for $\tilde{p}$ and $\tilde{q}$. Therefore we have $(\tilde{p}, \tilde{q}) \equiv (p_{\text{cone}}, q_{\text{cone}}) \equiv (0,0)$, and Proposition 2 is proved. □

5. Expanding Ricci solitons

Theorem 1. *Let $m, n \geq 2$ und $a \in \mathbb{R}$. There exist $b_1 = b_1(a) > 0$ and $\lambda_1, \lambda_2 > 0$ such that for any $b \in (0, b_1]$ the expanding soliton described by $(\varphi_{1,a,b}, \psi_{1,a,b})$ is complete and asymptotically looks like a cone over $\lambda_1 S^m \times \lambda_2 N$. This means that we have a family of complete expanding Ricci solitons on $\mathbb{R}^{m+1} \times N$, defined for all times $t > 0$ and emerging from a cone at $t = 0$.*

The proof of this theorem is based on the following lemma. Let $A(y) := |J - p'|(y)$, with $J = J(p, q; y)$ from Section 1, and $W(y)$ be as defined in (4).

Lemma 3. *Let $m, n \geq 2$, $a \in \mathbb{R}$ and $y_0 := \log(3(m+n)^2)$. Then the assumptions*

$$A(y_0) \leq \frac{9}{10}(m+n), \qquad W(y_0) \leq \frac{9}{25}$$

for any solution of (R3) imply that p, q, p' and q' stay bounded if $y \to \infty$. Moreover $\lim_{y\to\infty} p(y)$ and $\lim_{y\to\infty} q(y)$ exist.

Proof. From (5) and Young's inequality we obtain

$$W'(y) \leq (m+n)e^{-2y}A^2 - \left(m + n - 1 + \frac{1}{4}e^{2y} - A\right)\frac{p'^2 + mnq'^2}{m+n} \tag{20}$$

for any $y \in \mathbb{R}$. In addition we know from (6) that there holds

$$A'(y) \leq A + \sqrt{8(m+n)W} + 2W. \tag{21}$$

Step 1. *For any $y \geq y_0$ there holds:*

$$A(y) \leq \frac{e^{2y}}{5(m+n)^3}, \qquad W(y) \leq \frac{2e^{2y}}{25(m+n)^4}. \tag{22}$$

In any point $y \geq y_0$ satisfying the estimates (22) we deduce from (20) and (21)

$$W'(y) \leq \frac{e^{2y}}{25(m+n)^5} - \left(m + n - 1 + \frac{e^{2y}}{8}\right)\frac{p'^2 + mnq'^2}{m+n} \leq \frac{4e^{2y}}{25(m+n)^4} \tag{23}$$

and

$$A'(y) \leq \frac{e^{2y}}{5(m+n)^3} + \frac{4e^y}{5(m+n)^{3/2}} + \frac{4e^{2y}}{25(m+n)^4} \leq \frac{2e^{2y}}{5(m+n)^3}. \tag{24}$$

The assumptions of Lemma 3 show that inequality (22) holds strictly in $y = y_0$, which implies that before the first point with at least one equality in (22), either (23) or (24) must have been violated, hence there is no such first point. This proves the assertion of Step 1.

Step 2. *There exist a constant c_0 depending only on m and n such that for any $y \geq y_0$ there holds:*

$$A(y) \leq c_0 e^y, \qquad W(y) \leq W(y_0) + (m+n)c_0^2 y. \tag{25}$$

From (23) we infer

$$W'(y) \le \Big(\frac{1}{25(m+n)^5} - \frac{p'^2 + mnq'^2}{8(m+n)}\Big)e^{2y}$$

for $y \ge y_0$ which implies $(e^{-3y}W(y))' \le \Big(\frac{1}{25(m+n)^5} - \frac{p'^2+mnq'^2}{8(m+n)}\Big)e^{-y}$. Therefore we deduce (still for $y \ge y_0$)

$$\begin{aligned} 0 &\le e^{-3y}W(y) \le e^{-3y_0}W(y_0) + \int_{y_0}^{y} (e^{-3z}W(z))'\,dz \\ &\le \frac{2}{75(m+n)^6} + \int_{y_0}^{y} \Big(\frac{1}{25(m+n)^5} - \frac{p'^2 + mnq'^2}{8(m+n)}\Big)e^{-z}\,dz, \end{aligned}$$

which means

$$\begin{aligned} \int_{y_0}^{y} (p'^2 + mnq'^2)e^{-z}\,dz &\le \frac{16}{25(m+n)^3}\,e^{-y_0} + \frac{8}{25(m+n)^4}\int_{y_0}^{y} e^{-z}\,dz \\ &\le \Big(\frac{16}{25(m+n)^3} + \frac{8}{25(m+n)^4}\Big)e^{-y_0}. \end{aligned} \tag{26}$$

From (6), we have for any y

$$\{e^{-y}(J-p')\}' = e^{-y}\Big(2p' + \frac{p'^2 + mnq'^2}{m+n}\Big) \le e^{-y}(1 + 2(p'^2 + mnq'^2)),$$

and (26) shows for $y \ge y_0$

$$\begin{aligned} e^{-y}(J-p')(y) &\le e^{-y_0}(J-p')(y_0) + \int_{y_0}^{y} (1 + 2(p'^2 + mnq'^2))e^{-z}\,dz \\ &\le \frac{e^{y_0}}{10(m+n)^3} + \Big(1 + \frac{16}{25(m+n)^4} + \frac{32}{25(m+n)^3}\Big)e^{-y_0}. \end{aligned} \tag{27}$$

On the other hand, (6) also implies $(e^{-y}(J-p'))' \ge -(m+n)e^{-y}$, and we obtain the lower bound

$$e^{-y}(J-p')(y) \ge -\frac{e^{y_0}}{10(m+n)^3} - (m+n)e^{-y_0}. \tag{28}$$

Combining (27), (28) (and recalling $m+n \ge 4$) we deduce $A(y) \le c_0 e^y$ for $y \ge y_0$, where we can choose $c_0 := \frac{3}{10(m+n)} + \frac{1}{3(m+n)^2}$. Using this estimate for A we now obtain from (20)

$$W'(y) \le (m+n)e^{-2y}A^2 - \frac{p'^2 + mnq'^2}{8(m+n)}\,e^{2y} \le (m+n)c_0^2 - \frac{p'^2 + mnq'^2}{8(m+n)}\,e^{2y}, \tag{29}$$

for $y \ge y_0$, which gives the improved estimate for W in (25). Thus we have proved the assertions of Step 2. Now let $[z_1, z_2]$ be an interval in $[y_0, \infty)$ where $\frac{p'^2+mnq'^2}{m+n} \ge$

e^{-y}. From (29) and (25) we obtain

$$\begin{aligned} 0 &\leq W(z_2) \leq W(y_0) + (m+n)c_0^2 z_1 + \int_{z_1}^{z_2} \left((m+n)c_0^2 - \frac{e^y}{8(m+n)} \right) dy \\ &= \frac{1}{8(m+n)} \left(e^{z_1} - (1 - 8(m+n)(W(y_0) + (m+n)c_0^2 z_2) e^{-z_2}) e^{z_2} \right) \\ &\leq \frac{1}{8(m+n)} \left(e^{z_1} - e^{z_2 - c_1(z_2+1)e^{-z_2}} \right), \end{aligned}$$

where we have used $8(m+n)(W(y_0)+(m+n)c_0^2 z_2)e^{-z_2}) < \frac{1}{2}$ and the monotonicity of $z \mapsto e^{-z}z$. This estimate implies

$$z_2 - z_1 \leq c_1(z_2+1)e^{-z_2} \leq c_2 e^{-z_2/2}. \tag{30}$$

For $y \in [z_1, z_2]$ we further obtain via (25) (and $y \geq y_0 > 0$)

$$\left(\frac{p'^2 + mnq'^2}{2(m+n)} \right)' \leq W' - \frac{m+n-1}{2} \frac{d}{dy} \left(ne^{-\frac{2(p-mq)}{m+n}} + me^{-\frac{2(p+nq)}{m+n}} + 2p \right) \leq c_3 y^{3/2}.$$

From this estimate we deduce via (30) that there exist constants c_3, c_4 and c_5 depending only on n and m such that for any $y \in [z_1, z_2]$ there holds.

$$\begin{aligned} \frac{p'^2 + mnq'^2}{m+n} &\leq e^{-z_2} + c_3(z_2^{5/2} - z_1^{5/2}) \leq e^{-z_2} + c_4 z_2^{3/2}(z_2 - z_1) \\ &\leq e^{-z_2} + c_2 c_4 z_2^{3/2} e^{-z_2/2} \leq e^{-y} + c_5 e^{-y/4}. \end{aligned} \tag{31}$$

Note that the characterization of the intervals $[z_1, z_2]$ shows that this estimate also holds on the complement of such intervals in $[y_0, \infty)$. Therefore (31) holds for any $y \geq y_0$, and $|p'|$, $|q'|$ stay bounded as y tends to ∞. Furthermore, integration of the estimates for p' and q' (from the square root of (31)) shows boundedness $|p|$ und $|q|$, and further the existence of $p(\infty)$ and $q(\infty)$. □

Proof of Theorem 1. According to Proposition 2 for all b sufficiently small (depending on a) we have that $p_{1,a,b}, q_{1,a,b}$ satisfy the assumptions of Lemma 3. The assertion of this lemma implies the theorem. □

References

[AIV] S. Angenent, T. Ilmanen, J.J.L. Velazquez: *Fattening from smooth initial data in mean curvature flow.* In preparation, see [Il].

[B1] C. Böhm: *Inhomogeneous Einstein metrics on low-dimensional spheres and other low-dimensional spaces.* Invent. Math. **134** (1998), 145–176.

[B2] C. Böhm: *Non-compact cohomogeneity one Einstein manifolds.* Bull. Soc. Math. France **127** (1999), 135–177.

[EW] J. Eschenburg, M. Wang: *The initial value problem for cohomogeneity one Einstein manifolds.* J. Geom. Anal. **10** (2000), 109–137.

[FIK] M. Feldman, T. Ilmanen, D. Knopf: *Rotationally symmetric shrinking and expanding gradient Kähler-Ricci solitons.* Preprint, 2002.

[FK] D. Ferus, H. Karcher: *Non-rotational minimal spheres and minimizing cones.* Comm. Math. Helv. **60** (1985), 247–269.

[G] A. Gastel: *Nonuniqueness for the Yang-Mills heat flow.* J. Diff. Eq. **187** (2003), 391–411.

[H] R.S. Hamilton: *The formation of singularities in the Ricci flow.* Surveys in differential geometry, Vol. II (Cambridge, MA, 1993), 7–136, Internat. Press, Cambridge, MA, 1995.

[Il] T. Ilmanen: *Nonuniqueness in geometric heat flows.* In Lectures on mean curvature flow and related equations, Trieste (1995), lecture 4, 38–56. `http://www.math.ethz.ch/~ilmanen/notes.ps`

[Iv] T. Ivey: *New examples of complete Ricci solitons.* Proc. Amer. Math. Soc. **122** (1994), 241–245.

Andreas Gastel
Mathematisches Institut der
Heinrich-Heine-Universität Düsseldorf
Universitätsstraße 1
D-40225 Düsseldorf, Germany
e-mail: `gastel@cs.uni-duesseldorf.de`

Manfred Kronz
M.K.: Mathematisches Institut der
Universität Erlangen-Nürnberg
Bismarckstraße 1 1/2
D-91054 Erlangen, Germany
e-mail: `kronz@mi.uni-erlangen.de`

Progress in Nonlinear Differential Equations
and Their Applications, Vol. 59, 95–109

Evolution by Curvature of Networks of Curves in the Plane

Carlo Mantegazza – Joint Project with Matteo Novaga and Vincenzo Maria Tortorelli

Abstract. This survey describes our project to study the motion by curvature of a network of smooth curves with multiple junctions in the plane, that is, the geometric gradient flow associated to the *Length* functional.

Such a flow can represent the evolution of a two-dimensional multiphase system where the energy is simply the sum of the lengths of the interfaces, in particular it is a possible model for the growth of grain boundaries.

Moreover, the motion of these networks of curves is the simplest example of curvature flow for sets which are "essentially" non regular.

In this paper, we introduce the problem and we present some results and open problems about existence, uniqueness and, in particular, the global regularity of the flow.

1. Introduction

In this survey we describe the first steps of our project to address the problem of the motion by curvature of a network of curves in the plane, where by network of curves we mean a connected planar graph without self-intersections.

The evolution by curvature of such a network is the geometric gradient flow with respect to the energy given by the *Length functional* which is simply the sum of the lengths of all the curves of the network (see [7]).

We point out two motivations to study this evolution. The first is the analysis of models of two-dimensional multiphase systems, where the problem of the structure and regularity of the interfaces between different phases arises naturally. As an example, the model where the energy of a configuration is simply the total length has proven useful in the analysis of the growth of grain boundaries, see [7, 8, 15, 22], the papers by Herring and Mullins in [6] and *http://mimp.mems.cmu.edu*.

The second motivation is more theoretical: the evolution of such a network of curves in the plane is the simplest example of motion by mean curvature of a set which is *essentially* singular.

Mathematics Subject Classification (2000). Primary 53C44; Secondary 53A04, 35K55.
Key words and phrases. Curvature, evolution, triple junction.

In the literature there are various generalized definitions of flow by mean curvature for non regular sets (see [1, 7, 11, 13, 21, 25], for instance). All of them are fairly general, but usually lack uniqueness and a satisfactory regularity theory, even in simple situations.

We consider a connected network $\mathbb{S} = \cup_{i=1}^{n} \sigma^i$ in a smooth domain $\Omega \subset \mathbb{R}^2$ to be a finite family of regular curves $\sigma^i(x) : [0,1] \to \overline{\Omega}$ which can intersect each other or self-intersect only at their end points. We call "multi-points" the vertices of such a smooth graph $\mathbb{S}$ whose order is greater than one. Moreover, we assume that all the other ends of the curves (if there are any) are some fixed points P^l on the boundary of Ω.

The problem is to analyse the existence, uniqueness, regularity and asymptotic behaviour of the evolution by curvature of this network, under the constrain that the end points $P^l \in \partial\Omega$ stay fixed.

Inspired by Grayson's Theorem in [14], stating that any smooth closed curve embedded in $\mathbb{R}^2$ evolves by curvature without singularities before vanishing, and by the new approach to such result by Huisken in [20], one can reasonably expect that an "embedded" network of smooth curves does not develop singularities during the flow if its "topological structure" does not change (we will be more precise about this point in the sequel) and asymptotically converges to a critical configuration for the *Length* functional.

In this survey we present our first results, discussing some missing key points and open problems, and we suggest possible future research directions.

In many places this paper will be somehow roughly, we preferred to discuss mainly the general problems than to mention all the technical points.

We refer the interested reader to the paper [23] for proofs and all the details.

2. Basic definitions and small time existence of the flow

Since the sets we consider are singular because of the presence of the multi-points, first of all we need to decide what definition of flow by mean curvature we adopt.

As previously underlined, the existing weak definitions of curvature motion do not give uniqueness of the flow or allow "fattening" phenomena (see [13], for instance), which we would like to avoid. Among the existing notions, the most suitable to our point of view seems to be Brakke's one (see Definition 2.2), which also lacks uniqueness, but maintains at least the Hausdorff dimension of the sets (thus, preventing the event of fattening).

At the moment we are able to show a satisfactory *small time* existence result (Theorem 2.4) of a smooth motion for a special class of networks, that is, the ones having only multi-points with at most three concurrent curves (triple-points) forming angles of 120 degrees between them (this last property is called *Herring condition*).

We set in a precise analytical way the curvature evolution problem for an embedded special network with only triple-points.

Definition 2.1. We say that the family of networks of curves $\mathbb{S}_t = \cup_{i=1}^n \gamma^i(\cdot, t)$ in $\Omega \subset \mathbb{R}^2$, with end points $P^j \in \partial\Omega$ and only triple-points, evolves by curvature (remaining embedded) in the time interval $[0, T)$ if the functions $\gamma^i : [0,1] \times [0,T) \to \overline{\Omega}$ are of class C^2 in space and C^1 in time, at least, and satisfy the following quasilinear parabolic system

$$\begin{cases} \gamma_x^i(x,t) \neq 0 & \text{regularity} \\ \gamma^i(x,0) = \sigma^i(x) & \text{initial data} \\ \gamma_t^i(x,t) = \frac{\gamma_{xx}^i(x,t)}{|\gamma_x^i(x,t)|^2} & \text{motion by curvature} \end{cases} \tag{2.1}$$

for every $x \in [0,1]$, $t \in [0,T)$ and $i \in \{1, \dots, n\}$, moreover, the following conditions hold,

1. at every time, the curves can intersect each other or self-intersect only at their ends (embeddedness of the network);
2. every end of a curve, either is a 3-point or it coincides with one of the fixed end points of the network P^j on the boundary of Ω (there are no "free" end points of the curves). In this latter case, for example, if $\sigma^i(0) = P^j$, for some index j, then $\gamma^i(0,t) = P^j$ for every $t \in [0,T)$;
3. every three curves meeting at a 3-point of the network form three angles of 120 degrees;
4. the only curves which "touch" the boundary of Ω are the ones with fixed end points coinciding with the points P^j.

Notice that the evolution equation

$$\gamma_t^i = \frac{\gamma_{xx}^i}{|\gamma_x^i|^2}, \tag{2.2}$$

is not the usual way to describe the motion by curvature, that is,

$$\gamma_t^i = \frac{\langle \gamma_{xx}^i \,|\, \nu^i \rangle}{|\gamma_x^i|^2} \nu^i = k^i \nu^i$$

where we denoted with ν^i the unit normal to the curve γ^i and k^i its curvature.

The two velocities differ by a tangential component which actually affects the motions of the single points (Lagrangian point of view), but it does not affect the local motion of a curve as a whole subset of $\mathbb{R}^2$ (Eulerian point of view).

In our situation this extra component becomes necessary in order to allow the motion of the 3-points. Indeed, since we look for a C^2 solution of problem (2.1), if the velocity would be in normal direction at every point of three concurrent curves, the 3-point should move in a direction which is normal to all three, thus the only possibility would be that it does not move at all (see the discussions and examples in [7, 8, 22]).

Definition 2.2. We will speak of *Brakke flow with equality* of an initial network $\mathbb{S}_0$ in $[0,T)$, for a family of C^2 networks $\mathbb{S}_t$ in Ω, all with the same end points as $\mathbb{S}_0$

and satisfying the equation

$$\frac{d}{dt}\int_{\mathbb{S}_t} \varphi(\gamma,t)\,ds = -\int_{\mathbb{S}_t} \varphi(\gamma,t)k^2\,ds + \int_{\mathbb{S}_t} \langle \nabla\varphi(\gamma,t)\,|\,\underline{k}\rangle\,ds + \int_{\mathbb{S}_t} \varphi_t(\gamma,t)\,ds\,, \quad (2.3)$$

for every smooth function with compact support $\varphi : \Omega\times[0,T)\to\mathbb{R}$ and $t\in[0,T)$.

This means also that the time derivative at the hand side has to exist.

Notice that the right-hand side does not give any problem since the curves are at least C^2.

It is straightforward to check that a solution of problem (2.1) is also a smooth Brakke flow with equality.

Remark 2.3. The original definition of Brakke flow stated in [7, Section 3.3] allows equality (2.3) to be an inequality (and networks $\mathbb{S}_t$ to be one-dimensional countably rectifiable subsets of $\mathbb{R}^2$ with a distributional notion of curvature, called *varifolds*, see [24]), precisely,

$$\begin{aligned}\frac{\overline{d}}{dt}\int_{\mathbb{S}_t} \varphi(x,t)\,d\mathcal{H}^1(x) \leq -\int_{\mathbb{S}_t} \varphi(x,t)k^2\,d\mathcal{H}^1(x) + \int_{\mathbb{S}_t} \langle \nabla\varphi(x,t)\,|\,\underline{k}\rangle\,d\mathcal{H}^1(x)\\ + \int_{\mathbb{S}_t} \varphi_t(x,t)\,d\mathcal{H}^1(x)\,,\end{aligned}$$

must hold for every *positive* smooth function with compact support $\varphi : \Omega\times[0,T)\to\mathbb{R}$ and $t\in[0,T)$, where $\frac{\overline{d}}{dt}$ is the *upper* derivative (the $\overline{\lim}$ of the incremental ratios) and $\mathcal{H}^1$ is the Hausdorff one-dimensional measure in $\mathbb{R}^2$.

This weaker condition was introduced by Brakke in order to prove an existence result [7, Section 4.13] for a family of initial sets much wider than the networks of curves, but, on the other hand, it let open the possibility of instantaneous vanishing of some parts of the set.

Since for our networks we are able to show, via a different method, the existence of a Brakke flow composed of smooth curves and satisfying the equality, for sake of simplicity, we included such extra properties in the definition.

Theorem 2.4. *If $\mathbb{S}_0$ is a C^2 initial embedded network, then there exists a Brakke flow with equality $\mathbb{S}_t$ for some positive time interval $[0,T)$.*

Moreover, the networks $\mathbb{S}_t$ it is described by a smooth family of curves solving problem (2.1) in every time interval $[\varepsilon,T)$, for any $\varepsilon>0$. Hence, all the curvatures k^i belong to $C^\infty([0,1]\times[\varepsilon,T))$, hence the flow is a smooth *Brakke flow with equality for every positive time.*

Finally, the unit tangents τ^i are continuous in space and time and the function $\int_{\mathbb{S}_t} k^2\,ds$ is continuous on $[0,T)$, where ds is the arclength measure on the curves of the network.

Remark 2.5. The fact that we have a solution of problem (2.1) (which is stronger than a Brakke flow) only after some positive time ε, is due to the eventuality that the *parabolic compatibility conditions* at time $t=0$ are not satisfied by the initial network. If such conditions hold, we have a *unique* solution of system (2.1) on all $[0, T)$, for every parametrization of the initial network $\mathbb{S}_0$ (see [23] for details).

The proof of the theorem is based on a result of Bronsard and Reitich in their paper [8], where they show that if the parabolic compatibility conditions are satisfied by the initial network, then the system (2.1) has a *unique* solution on some positive time interval $[0, T)$. Then, by means of a priori estimates and approximation, it is possible to get the existence of a Brakke flow for any C^2 network like above.

If the compatibility conditions are not satisfied, the uniqueness of a smooth Brakke flow with equality for these special networks, is an open problem, even restricting the admissible class of Brakke flow to the ones with continuous unit tangents.

We discuss now a little the difficulties in extending our analysis to a *general* network.

In the case of the presence in the initial network of a "bad" 3-point, not satisfying the Herring condition, that is, the three concurrent curves do not form angles of 120 degrees, we are not able at the moment to show the existence of a flow, smooth for every positive time, satisfying a "robust" definition (at least as Definition 2.2).

Actually, one would expect that the desired good definition should give uniqueness of the motion and force, by an instantaneous regularization, the three angles to become immediately of 120 degrees and to remain so. This is sustained by the fact that, by an energy argument ([7]), any smooth Brakke flow has to share such a property (which is also suggested by numerical and physical experiments, see at *http://mimp.mems.cmu.edu* and also the discussions in [6, 7, 8, 15, 22]).

Notice that, by the variational nature of the problem it is appealing to guess that some sort of parabolic regularization could play a role here.

The second "big trouble" is when one tries to consider networks with multi-points of order greater than three.

We remark that if a multi-point has only two concurrent curves, it can be shown, by the regularizing effect of the evolution by curvature (see [2, 4, 5, 14]), that the two curves together become instantaneously a single smooth curve moving by curvature. Hence, the 2-point has vanished but this particular event is so "soft" (and topologically null) that we can avoid to consider it as a real structural change.

In the case of a 4-point instead (and clearly also of a higher order multi-point), for instance considering the network described by two curves crossing each other, there are really several possible candidates for the flow, even excluding a priori "fattening" phenomena. One cannot easily decide how the angles must behave, like in the 3-point case above, moreover, one can allow the four concurrent curves to separate in two pairs of curves moving independently each other and it could even be taken into account the "creation" of new multi-points from such a single one (these events are actually possible in Brakke's definition).

In these latter cases, the topology of the network changes dramatically, forcing us to change the "structure" of the system (2.1) governing the evolution or the family of curves composing the network.

Finally, it should be noticed that, the previous situation can appear even for our special networks with only 3-points after some time, since two (or many) of them could possibly "collapse" together creating a 4-point (or a multi-point), modifying the topological structure of the network.

This clearly happens when the length of at least one curve of the network goes to zero (and there is no reason to exclude such an event). In this case, like at the previous point, one possibly has to "restart" the evolution with a different set of curves.

Again, a "collapse" could also produce a "bad" 3-points (think of three 3-points collapsing together along three curves connecting them).

Anyway, it seems actually reasonable that the configurations with multi-points of order greater than three or 3-points with angles different to 120 degrees should be unstable (they are actually unstable for the *Length* functional), with the meaning that they can appear at some discrete set of times (and probably in some cases are unavoidable), but they must vanish immediately after.

3. A priori estimates

Once the flow starts, by Theorem 2.4, we can suppose that it is described by a family of smooth curves $\gamma^i : [0,1] \times [0,T) \to \overline{\Omega}$, solving problem (2.1) in a *maximal time interval* $[0,T)$, hence, moving according to the law

$$\gamma^i_t(x,t) = \frac{\gamma^i_{xx}(x,t)}{|\gamma^i_x(x,t)|^2}\,.$$

After some time, it could happen that the network "hits" the boundary of Ω, so the flow would stop, by the definition above, without developing a *real* analytical singularity in the maps γ^i. Excluding for the moment this *geometric* event which will be discussed in the next section, there can be only two reasons why the time T is maximal: either the length of one curve γ^i has gone to zero (putting us in one of the situations discussed in the previous section) which implies that $\gamma^i_x = 0$, or some derivative of a curve γ^i is not bounded.

Hence, in order to analyse the long term behaviour, a priori estimates are needed. They are strongly based on several relations holding at the 3-points of the network $\mathbb{S}_t$ between the curvatures of the three concurrent curves and their space derivatives. For instance, the sum of the three curvatures at every 3-point is zero at every time.

Moreover, even if these 3-points (by the Herring condition) are *interior points* of the network, from the *distributional* point of view, every argument based on the maximum principle has to consider them as "genuine" *boundary points*. This fact forces us to change approach, with respect to the standard smooth case, and to resort to integral estimates obtained by means of interpolation inequalities on the

L^2 norms of the curvature and its derivatives. Pointwise estimates then will follow using Sobolev embeddings.

The first, immediate, estimate is on the lengths of the curves: since we are moving by the gradient of the *Length* functional, the length of every curve of the network is uniformly bounded above by a constant independent of time.

After some heavy computation one is able to get the following proposition which control these lengths also away from zero.

Proposition 3.1. *For every $M > 0$ there exists a time $T_M \in (0, T)$ such that for every $\mu > 0$, the inverses of the lengths of the curves of $\mathbb{S}_0$, the curvature of every $\mathbb{S}_t$ and all its space derivatives are uniformly bounded in the time interval $[\mu, T_M]$ by some constants depending only on μ, the L^2 norm of the curvature of $\mathbb{S}_0$ and the inverses of the lengths of the curves of the network at time zero.*

By means of these a priori estimates, which are similar to the ones in [3, 4, 5, 18] for the mean curvature evolution of a smooth curve or hypersurface, we can work out some results about the flow and, in particular, what happens at the maximal time of existence.

Theorem 3.2. *If $[0, T)$ is the maximal time interval of existence of a smooth solution $\mathbb{S}_t$ with $T < +\infty$ of problem (2.1), then*

1. *either the inferior limit of the length of at least one curve of $\mathbb{S}_t$ goes to zero as $t \to T$,*
2. *or $\overline{\lim}_{t\to T} \max_{\mathbb{S}_t} k^2\, ds = +\infty$.*

Moreover, if the lengths of the curves are uniformly bounded away from zero, then the superior limit in (2) is a limit and there exists a positive constant C such that

$$\max_{\mathbb{S}_t} k^2 \geq \frac{C}{\sqrt{T-t}} \to +\infty\,, \tag{3.1}$$

for every $t \in [0, T)$.

Remark 3.3. In the case of the evolution γ_t of a single closed curve in the plane there exist a constant $C > 0$ such that if at time $T > 0$ a singularity develops, then

$$\max_{\gamma_t} k^2 \geq \frac{C}{T-t}$$

for every $t \in [0, T)$ (see [19]). If this lower bound on the rate of blowing up of the curvature (which is clearly stronger than the one in inequality (3.1)) holds also in the case of the evolution of a network is an open problem.

If we now suppose that no curve of the network collapses, the analysis proceed like in the standard case of mean curvature evolution of an hypersurface in $\mathbb{R}^n$.

What is needed is the understanding of the structure of the possible *blow up* around the singularities, in order to actually exclude these latter by means of geometric arguments. Some key references for this line are [3, 16, 19, 20].

The most relevant difference between our case and the standard one is that, because of the presence of the multi-points, it is difficult to apply the *maximum principle*, which would be the natural tool to get estimates on the geometric quantities during the flow.

We postpone the study of the possible singularities to Section 5, as we want now to discuss some geometric properties of the flow.

4. Geometric properties of the flow

As we said in the previous section, there are two "catastrophic" geometric events that do not allow the solution of system (2.1) to be defined for all times, one is the collapse of a curve that, at the moment, even requires a definition to be dealt with, the other is if the network during the flow "hits" the boundary of the domain $\Omega \in \mathbb{R}^2$.

Assuming the hypothesis that all the curves have lengths uniformly bounded away from zero, a simple condition to avoid the second situation is the convexity of Ω.

Proposition 4.1. *All the networks $\mathbb{S}_t$ intersect the boundary of Ω only at their end points.*

Moreover, for every positive time such intersections are transversal.

Another important geometric property of the flow, in particular if one is thinking to modelize a physical multiphase system in $\mathbb{R}^2$, for instance a situation with several non-mixing liquid, is the *embeddedness* of the network.

We required that all networks are embedded (condition (1) in problem (2.1)) but actually, one can get a solution for small time also starting from a self-intersecting network.

Clearly, if the initial network is embedded, by continuity, it remains so for some time, what is more interesting is that actually it remains embedded *for every time*, till the solution γ^i is smooth. Moreover, this geometric property plays a key role also in the analysis of the singularities.

We assume a couple of technical hypotheses, that we would like to drop in the future.

- The domain Ω is *strictly* convex.
- The network $\mathbb{S}_0$ is a *tree*, that is, it does not contain *loops*.

Borrowing ideas from [17], [20] (see also [10]), we associate to the flow $\mathbb{S}_t$ a quantity $E(t)$ as follows: considering two points p, q belonging to $\mathbb{S}_t$, we define $\Gamma_{p,q}$ to be the *geodesic* curve contained in $\mathbb{S}_t$ connecting p and q (such *geodesic* is uniquely defined because there are no loops in $\mathbb{S}_t$), then we let $A_{p,q}$ to be the area of the open region $\mathcal{A}_{p,q}$ in $\mathbb{R}^2$ enclosed by the segment $[p, q]$ and the curve $\Gamma_{p,q}$. When the region $\mathcal{A}_{p,q}$ is not connected, we let $A_{p,q}$ to be the sum of the areas of its connected components.

We consider the function $\Phi_t : \mathbb{S}_t \times \mathbb{S}_t \to \mathbb{R} \cup \{+\infty\}$ as

$$\Phi_t(p,q) = \begin{cases} \frac{|p-q|^2}{A_{p,q}} & \text{if } p \neq q, \\ 4\sqrt{3} & \text{if } p \text{ and } q \text{ coincide with a 3-point of } \mathbb{S}_t, \\ +\infty & \text{otherwise.} \end{cases}$$

Since $\mathbb{S}_t$ is smooth and the 120 degrees condition holds, it is easy to check that Φ_t is a lower semicontinuous function. Hence, by the compactness of $\mathbb{S}_t$, the following infimum is actually a minimum

$$E(t) = \inf_{p,q \in \mathbb{S}_t} \Phi_t(p,q) \tag{4.1}$$

for every $t \in [0,T)$.

If the network $\mathbb{S}_t$ has no self-intersections we have $E(t) > 0$, the converse is clearly also true.

Moreover, $E(t) \leq 4\sqrt{3}$ always holds, since at least one 3-point is present in the network, thus, when $E(t) > 0$ the two points (p,q) of a minimizing pair can coincide if and only if they coincide with a 3-point.

Finally, since the evolution is smooth it is easy to see that the function $E : [0,T) \to \mathbb{R}$ is continuous.

The nice property of this function E is stated in the following proposition.

Proposition 4.2. *The function $E(t)$ is monotone increasing in every time interval where $0 < E(t) < 4\sqrt{3}$ and for at least one minimizing pair (p,q) of Φ_t neither p nor q coincides with one of the end points P^i.*

Since the positivity of $E(t)$ is equivalent to the embeddedness of $\mathbb{S}_t$, our claim above follows.

We resume then these geometric properties.

Theorem 4.3. *If Ω is bounded and* strictly *convex, and $\mathbb{S}_0$ is a* tree, *there exists a constant $C > 0$ depending only on $\mathbb{S}_0$ such that $E(t) > C > 0$ for every $t \in [0,T)$.*

Hence, if the initial network was embedded, all the networks $\mathbb{S}_t$ remain embedded in all the maximal interval of existence of the flow.

We remark that the two hypotheses above seems to be really unnecessary and very likely can be avoided, maybe with a smart choice of a slightly different function E, since the property to be a tree is necessary only to have a *single* geodesic in order to well determine the region $\mathcal{A}_{p,q}$. Moreover, we remark that if one is able to extend this theorem to networks with loops, the strict convexity of Ω is no more necessary.

5. Analysis of the singularities

We recall that we continue to assume that the lengths of the curves of the networks are uniformly bounded away from zero.

Since at the maximal time T the curvature has to explode, like in the standard smooth case we separate the singularities according to its rate of blow up.

We say that a singularity is of *Type I* if for some constant C we have $\max_{\mathbb{S}_t} k^2 \leq C/(T-t)$ as $t \to T$ and it is of *Type II* otherwise.

Following Huisken [19], rescaling properly the flow around a "hypothetical" *Type I* singularity, one gets an evolution of embedded networks (unbounded and without end points) shrinking homothetically during the motion by curvature. Classifying all such particular evolutions, we will show that none of them can arise as a blow up of the flow $\mathbb{S}_t$, this clearly implies that *Type I* singularities cannot develop.

With the same idea, rescaling the flow around a *Type II* singularity, one gets an *eternal* motion by curvature, that is, an evolution of networks defined for every time $t \in \mathbb{R}$.

What is missing at the moment is that this eternal flow is actually simply given by a *translating* network (unbounded and without end points), like it happens in the case of a single smooth curve. Here also, the main difficulty resides in replacing some maximum principle arguments.

If this blow up would be translating, after classifying these latters, we could exclude also this case by means of an argument based on the monotonicity of the quantity E of previous section. Hence, no singularity at all could appear during the flow if the lengths of the curves of the network stay away from zero.

Since the technical details related to the blow up procedures are quite heavy, in this section we will speak a little bit roughly, referring to [23] for more precise statements.

Proposition 5.1. *Blowing up "in a proper way" the flow of the networks $\mathbb{S}_t$, in space and time, around a* Type I *singularity, we can obtain a nonempty "limit" flow by curvature of embedded smooth networks of curves in all $\mathbb{R}^2$, with at most one end point or 3-point (still satisfying the 120 degrees condition), which move simply by homothety with respect to the origin. Moreover, the curvature of these networks is not identically zero.*

At this point we want to know what are the possible candidate flows satisfying all the properties stated in this proposition.

Lemma 5.2. *The only smooth flows by curvature of embedded networks with at most one end point or 3-point, moving by homothety with respect to the origin of $\mathbb{R}^2$ are:*

1. *a straight line from the origin;*
2. *a halfline from the origin;*
3. *the network composed by three halfline from the origin, forming three angles of 120 degrees between them.*

Since in all these three situations the curvature of the network is identically zero, we get a contradiction.

Theorem 5.3. Type I *singularities cannot develop during the smooth flow* $\mathbb{S}_t$ *of an embedded network in a bounded and strictly convex domain* $\Omega \subset \mathbb{R}^2$ *if the lengths of the curves are uniformly bounded away from zero.*

We turn now our attention to the *Type II* singularities, that is,

$$\overline{\lim_{t \to T}} (T - t) \max_{\mathbb{S}_t} k^2 = +\infty\,.$$

Employing a trick by Hamilton, blowing up the flow in a way different to the previous one, we get an *eternal* limit flow by curvature of embedded smooth networks $\mathbb{S}_t^\infty$, all with at most one end point or 3-point (as before), where *eternal* means that the flow is defined for every time $t \in \mathbb{R}$. Moreover, the curvature of the networks achieves a non zero maximum at a point in space and time.

In the case of the evolution of a single closed curve in the plane, it is possible to show that the "limit" flow by curvature obtained in this way has to move by translation. This conclusion is reached in two steps: first, one shows that at every time the limit curve is convex, then, by means of the Harnack estimate proved by Hamilton in [16], it follows that it is simply translating during the motion.

Then, one is able to classify these translating flows and it turns out that all of them are, up to translations and dilations, the so called *grim reaper* (see [3, 16]), which is the curve given by the graph of the function $x = -\log(\cos y)$ in $\mathbb{R}^2$ when y varies in the interval $(-\pi/2, \pi/2)$. Notice that this curve is contained in a strip

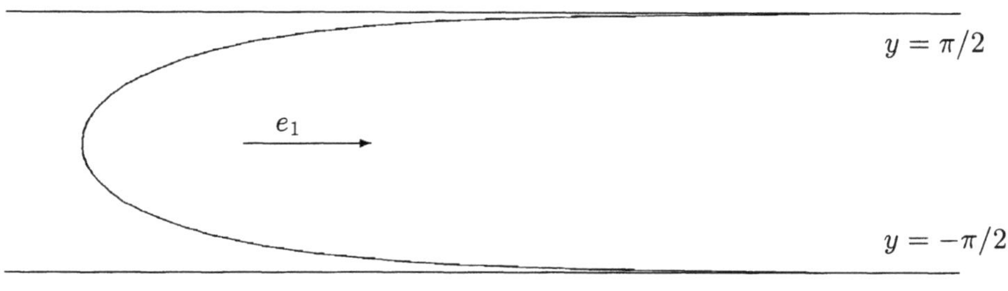

FIGURE 1. The *grim reaper*.

or $\mathbb{R}^2$ of finite width.

In our situation the convexity condition above can be translated to "the curvature of the limit flow is never zero".

At the moment we can only state the following two conjectures.

Conjecture 5.4. If the curvature is zero at some point of the limit flow of networks (or curve) $\mathbb{S}_t^\infty$, then it is zero along all the curve containing such a point (everywhere).

Conjecture 5.5. The networks (or curves) $\mathbb{S}_t^\infty$ move by translation.

If this last conjecture is true, so $\mathbb{S}_t^\infty$ are translating networks, we can apply the following classification lemma to describe them.

Lemma 5.6. *The only smooth flows by curvature of embedded networks with at most one end point or* 3*-point, moving by translation in* $\mathbb{R}^2$ *are:*

1. *a straight line (not moving at all);*
2. *a halfline;*
3. *a* triod *composed by three halflines (like the set at point* (3) *of Lemma* 5.2*;*
4. *the* grim reaper*;*
5. *a network composed of three unbounded curves which are either halflines or translated copies of pieces of the* grim reaper*, meeting at a single* 3*-points with angles of* 120 *degrees (it clearly follows that at most one curve can be a halfline).*

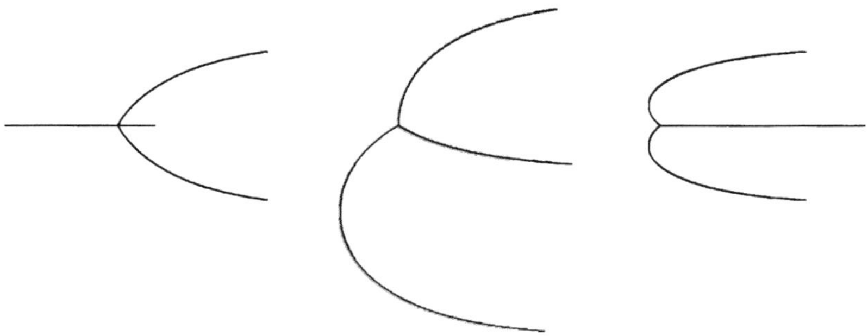

FIGURE 2. Some examples of translating networks.

Since our limit flow has non zero curvature, the first three cases would be excluded and we could conclude that either we are dealing with the *grim reaper* or with a network composed by three pieces of it, or two pieces and a halfline.

If we can get to this conclusion, we consider the analogous of the function E of Section 4 for the limit flow $\mathbb{S}_{\mathfrak{t}}^\infty$ (defined on all $\mathbb{R}$ in this case), we call it E^∞ to distinguish it from the one of the original flow $\mathbb{S}_t$. Since, by definition, E is dilation invariant and we got these limit flows by means of a blow up procedure, as we know that $E(t)$ was greater than an universal constant C (by the embeddedness of the initial network), the same estimate holds also for E^∞, that is, $E^\infty(\mathfrak{t}) \geq C$ for every $\mathfrak{t} \in \mathbb{R}$.

Fixing $t \in \mathbb{R}$, if we now consider a pair of points p, q on two curves of the network with opposite convexity (or on the two ends, if $\mathbb{S}_{\mathfrak{t}}^\infty$ is the *grim reaper*), we see that sending them both to infinity in a way that the length of the segment $[p, q]$ is uniformly bounded (this can be done because the *grim reaper* is contained in a strip of $\mathbb{R}^2$ with finite width), it follows that $E^\infty(\mathfrak{t})$ must be zero, which is in contradiction with $E^\infty(\mathfrak{t}) \geq C$.

This clearly exclude *Type II* singularities.

Remark 5.7. The same argument, but with a different geometric quantity, is used by Huisken in [20] to exclude *Type II* singularities during the motion of a single curve.

Proposition 5.8. *If Conjecture* 5.5 *is true then* Type II *singularities cannot develop during the smooth flow* $\mathbb{S}_t$ *of an embedded network in a bounded and strictly convex domain* $\Omega \subset \mathbb{R}^2$*, if the lengths of the curves are uniformly bounded away from zero.*

Hamilton's proof that a convex blow up of a *Type II* singularity (in the standard smooth case) is translating, is heavily based on the maximum principle which, as we said, is difficult to apply in our situation. So, it could happen that only Conjecture 5.4 can be proved and one could possibly exclude *Type II* singularities without actually show that the limit flow $\mathbb{S}^\infty_t$ is translating. For instance, if the curvature is always or never zero on each of the three curves, then they have asymptotic tangents (indeed, they are all convex but some of them in the opposite way, by the fact that the sum of the curvatures at the 3-point is zero), hence in order to apply the previous argument based on the function E, it would be enough to show that two of these limit tangents, belonging to curves with opposite convexity, coincide.

Conjecture 5.9. The flow $\mathbb{S}_t$ of an embedded network in a bounded and strictly convex domain $\Omega \subset \mathbb{R}^2$, such that the lengths of its curves are uniformly bounded away from zero, does not develop singularities at all.

If this conjecture is true, it can also be shown that the networks converge, as $t \to +\infty$ to a configuration which is critical for the *Length* functional among all the networks Ω with the same end points.

6. Open problems

The main problem left open in the paper is Conjecture 5.5, whose validity would imply Conjectures 5.9.

Other questions in the paper that we would like to set are concerned with the extension of the results to all the networks, possibly with loops, with only 3-points (no 4-points or higher order points), in particular, proving an analogous of Theorem 4.3. This would also make superfluous the requirement that the domain $\Omega \subset \mathbb{R}^2$ is *strictly* convex.

Finally, we conclude by listing some, naturally arising, research directions.

1. The problem of the uniqueness of Brakke flows (smooth/with equality) of networks, discussed after Theorem 2.4.
2. The problem of the existence/uniqueness of flow for an initial network not satisfying the 120 degrees condition at the 3-points.
3. The study of the singularities such that the curvature blows up but the lengths are not bounded away from zero. Such analysis requires new estimates and a general classification of all homothetically moving and translating networks, with only 3-points, flowing by curvature.
4. The "definitory" problem of the motion of networks with multi-points (of order greater than three) and the analysis of the collapsing situations with change of topology (see the introduction and the papers by De Giorgi [12] and Caraballo [9]).

References

[1] F.J. Almgren, J.E. Taylor, and L. Wang, *Curvature driven flows: a variational approach*, SIAM J. Cont. and Opt. **31** (1993), 387–438.

[2] S. Altschuler and M. Grayson, *Shortening space curves and flow through singularities*, J. Diff. Geom. **35** (1992), 283–298.

[3] S.J. Altschuler, *Singularities of the curve shrinking flow for space curves*, J. Diff. Geom. **34** (1991), no. 2, 491–514.

[4] S. Angenent, *Parabolic equations for curves on surfaces: curves with p-integrable curvature*, Ann. of Math. (2) **132** (1990), 451–483.

[5] ———, *Parabolic equations for curves on surfaces: intersections, blow up and generalized solutions*, Ann. of Math. (2) **133** (1991), 171–215.

[6] J.M. Ball, D. Kinderlehrer, P. Podio-Guidugli, and M. Slemrod (eds.), *Fundamental Contributions to the Continuum Theory of Evolving Phase Interfaces in Solids*, Springer, Berlin, 1999.

[7] K.A. Brakke, *The Motion of a Surface by its Mean Curvature*, Princeton University Press, Princeton, N.J., 1978.

[8] L. Bronsard and F. Reitich, *On three-phase boundary motion and the singular limit of a vector-valued Ginzburg-Landau equation*, Arch. Rat. Mech. Anal. **124** (1993), no. 4, 355–379.

[9] D.G. Caraballo, *A variational scheme for the evolution of polycrystals by curvature*, Ph.D. thesis, Princeton University, 1996.

[10] K.-S. Chou and X.-P. Zhu, *Shortening complete plane curves*, J. Diff. Geom. **50** (1998), no. 3, 471–504.

[11] E. De Giorgi, *Barriers, boundaries and motion of manifolds*, Scuola Normale Superiore di Pisa, 1995.

[12] ———, *Motions of partitions*, Variational methods for discontinuous structures (Como, 1994), Progr. Nonlinear Differential Equations Appl., vol. 25, Birkhäuser, Basel, 1996, pp. 1–5.

[13] L.C. Evans and J. Spruck, *Motion of level sets by mean curvature I*, J. Diff. Geom. **33** (1991), 635–681.

[14] M.A. Grayson, *The heat equation shrinks embedded plane curves to round points*, J. Diff. Geom. **26** (1987), 285–314.

[15] M.E. Gurtin, *Thermomechanics of Evolving Phase Boundaries in the Plane*, Oxford Science Publication, New York, 1993.

[16] R.S. Hamilton, *Harnack estimate for the mean curvature flow*, J. Diff. Geom. **41** (1995), no. 1, 215–226.

[17] ———, *Isoperimetric estimates for the curve shrinking flow in the plane*, Modern methods in complex analysis (Princeton, NJ, 1992), Princeton University Press, Princeton, NJ, 1995, pp. 201–222.

[18] G. Huisken, *Flow by mean curvature of convex surfaces into spheres*, J. Diff. Geom. **20** (1984), 237–266.

[19] ———, *Asymptotic behavior for singularities of the mean curvature flow*, J. Diff. Geom. **31** (1990), 285–299.

[20] ———, *A distance comparison principle for evolving curves*, Asian J. Math. **2** (1998), 127–133.

[21] T. Ilmanen, *Elliptic regularization and partial regularity for motion by mean curvature*, Mem. Amer. Math. Soc., vol. 108(520), AMS, 1994.

[22] D. Kinderlehrer and C. Liu, *Evolution of grain boundaries*, Math. Models Methods Appl. Sci. **11** (2001), no. 4, 713–729.

[23] C. Mantegazza, M. Novaga and V.M. Tortorelli, *Motion by curvature of planar networks*, in preparation.

[24] L. Simon, *Lectures on Geometric Measure Theory*, Proc. Center Math. Anal., vol. 3, Australian National University, Canberra, 1983.

[25] H.M. Soner, *Motion of a set by the curvature of its boundary*, J. Diff. Eqs. **101** (1993), no. 2, 313–372.

Carlo Mantegazza
Scuola Normale Superiore di Pisa
P.za Cavalieri 7
I-56126, Pisa, Italy
e-mail: `mantegazza@sns.it`

Matteo Novaga
Math. Dept. Univ. Pisa
I-56127 Pisa, Italy
e-mail: `novaga@dm.unipi.it`

Vincenzo Maria Tortorelli
Math. Dept. Univ. Pisa
I-56127 Pisa, Italy
e-mail: `tortorel@dm.unipi.it`

Part III

Harmonic Mappings in Special Geometries

Progress in Nonlinear Differential Equations
and Their Applications, Vol. 59, 113–132

Harmonic Maps in Complex Finsler Geometry

Seiki Nishikawa

Dedicated to Professor Takushiro Ochiai on his sixtieth birthday

Abstract. Given a smooth map from a compact Riemann surface to a complex manifold M equipped with a strongly pseudoconvex Finsler metric F, we define a natural notion of the $\bar{\partial}$-energy of the map. A harmonic map is then defined to be a critical point of the $\bar{\partial}$-energy functional. Under the condition that F is weakly Kähler, we obtain the second variation formula of the functional, and prove that any $\bar{\partial}$-energy minimizing harmonic map from a Riemann sphere to a weakly Kähler Finsler manifold M of positive curvature is either holomorphic or antiholomorphic. Significance of complex Finsler metrics in the study of holomorphic vector bundles, in particular its relation with the Hartshorne conjecture in complex algebraic geometry, is represented. A brief overview of complex Finsler geometry is also provided.

1. Introduction

In 1974 Kobayashi [12] proved, in his address at Oberwolfach, that a holomorphic vector bundle E over a compact complex manifold M is negative, that is, its dual vector bundle E^* is ample, if and only if it admits a strongly pseudoconvex Finsler metric F with negative curvature. This differential geometric characterization of negative or ample vector bundles, in terms of complex Finsler metrics, had paved a new way for introducing differential geometric techniques into the study of ample and negative vector bundles.

Since then, complex Finsler metrics have attracted much renewed attention to possible applications in various fields in complex geometry. In particular, in order to investigate intrinsic metrics on complex manifolds, the notion of complex geodesics was introduced by Vesentini [25] for the Carathéodory and the Kobayashi metrics, which are (in general not smooth) complex Finsler metrics. Roughly speaking, a complex geodesic is a holomorphic embedding of the unit disk in the complex plane, equipped with the Poincaré metric, which is isometric with respect to the intrinsic metric defined on the manifold under consideration. The question of the existence and uniqueness of these complex geodesics is natu-

---

Mathematics Subject Classification (2000). Primary 53C43; Secondary 53B40, 53C55.
Key words and phrases. Harmonic maps, holomorphic maps, complex Finsler geometry, weakly Kähler Finsler metric, Hartshorne conjecture, Frankel conjecture.
* Partly supported by the Grant-in-Aid for Scientific Research (A), No. 15204003, and the Grant-in-Aid for Exploratory Research, No. 14654011, of the Japan Society for the Promotion of Science.

rally related to the existence of "Kähler" Finsler metrics of constant holomorphic curvature. This latter fact has eventually led Abate and Patrizio [2, 3] to a classification of Kähler Finsler manifolds of constant holomorphic curvature under some appropriate condition on the symmetry of curvature.

In spite of these satisfactory results, one might recognize that a somewhat more natural geometric object has been overlooked for a long time, namely, the notion of harmonic maps from a compact Riemann surface into a complex Finsler manifold, which may be considered as a proper generalization of complex geodesics and expected to be more flexible than them.

The purpose of the present paper is to define, as the first step in initiating the study in this direction, a natural notion of the $\bar{\partial}$-energy $E_{\bar{\partial}}(f)$ for a given smooth map $f : N \to M$ from a compact Riemann surface N to a complex manifold M equipped with a strongly pseudoconvex Finsler metric F. A harmonic map is then defined to be a critical point of the $\bar{\partial}$-energy $E_{\bar{\partial}}(f)$, considered as a functional defined on the space of smooth maps from N to M. The first variation formula of this $\bar{\partial}$-energy functional turns out to involve a certain Kählerianity condition on the Finsler metric F, which is called "weakly Kähler" in Abate and Patrizio [1]. Under the condition that F satisfies this weakly Kähler condition, which assumes the vanishing of certain contraction of the torsion of the canonical connection defined by F, we obtain in Section 5 the equations defining harmonic maps in this context, which reduces to the well-known harmonic map equation provided F is a usual Kähler metric. Then in Section 7, under the same Kählerianity condition, we obtain the second variation formula of the $\bar{\partial}$-energy functional, and prove the following complex analyticity of $\bar{\partial}$-energy minimizing harmonic maps as its application.

Theorem. *Let M be a complex manifold equipped with a weakly Kähler Finsler metric F which has positive curvature. If $f : P_1(\boldsymbol{C}) \to M$ is a $\bar{\partial}$-energy minimizing harmonic map from a Riemann sphere $P_1(\boldsymbol{C})$ to M, then f is either holomorphic or antiholomorphic.*

The author is grateful to Professors Sorin Dragomir and Atsushi Tachikawa for their useful discussions during the preparation of the paper.

2. Complex Finsler metrics

Let E be a holomorphic vector bundle of rank r over a complex manifold M of dimension m with projection π. We denote a point of E by (z, ζ), where z represents a point of M and ζ is a vector in the fibre $E_z = \pi^{-1}(z)$ of E over $z \in M$. Let $o : M \to E$ be the zero section of E and set $E^\circ = E \setminus o(M)$, that is, E° denotes E minus its zero section. A *complex Finsler metric* F on E is a real function $F : E \to \boldsymbol{R}$ on E satisfying the following conditions:

(i) $F(z, \zeta) \geq 0$, and $F(z, \zeta) = 0$ if and only if $\zeta = 0$.
(ii) $F \in C^\infty(E^\circ)$, that is, F is smooth outside the zero section of E.
(iii) $F(z, \lambda\zeta) = |\lambda| F(z, \zeta)$ for all $(z, \zeta) \in E$ and $\lambda \in \boldsymbol{C}$.

In the following, we set $G(z,\zeta) = F^2(z,\zeta)$, since we use $F^2(z,\zeta)$ more often than $F(z,\zeta)$. Let $z = (z^1,\dots,z^m)$ be a local coordinate system for M and $\zeta = (\zeta^1,\dots,\zeta^r)$ the local fibre coordinate system defined by a local holomorphic frame field $s = (s_1,\dots,s_r)$ of E. Then $(z,\zeta) = (z^1,\dots,z^m,\zeta^1,\dots,\zeta^r)$ is a local coordinate system for E. We say that a complex Finsler metric F is *strongly pseudoconvex* if the complex Hessian

$$(G_{i\bar{j}}) = \left(\frac{\partial^2 G}{\partial\zeta^i\partial\bar{\zeta}^j}\right)$$

of G is positive definite on E°. For instance, if

$$g(z,\zeta) = \sum_{i,j=1}^{r} g_{i\bar{j}}(z)\zeta^i\bar{\zeta}^j \tag{2.1}$$

is a Hermitian metric on E, then $F(z,\zeta) = \sqrt{g(z,\zeta)}$ defines a strongly pseudoconvex Finsler metric on E.

The main property of G is its $(1,1)$-homogeneity in the sense that

$$G(z,\lambda\zeta) = \lambda\bar{\lambda}G(z,\zeta) \tag{2.2}$$

for all $(z,\zeta) \in E$ and $\lambda \in \boldsymbol{C}$. The significance of this condition in complex geometry is explained, for instance, in Rund [22]. In order to see the consequences of (2.2), we write

$$G_i = \partial G/\partial\zeta^i, \quad G_{\bar{j}} = \partial G/\partial\bar{\zeta}^j, \quad G_{i\bar{j}} = \partial^2 G/\partial\zeta^i\partial\bar{\zeta}^j,$$
$$G_{i\alpha} = \partial^2 G/\partial\zeta^i\partial z^\alpha, \quad G_{i\bar{j}\bar{\beta}} = \partial G_{i\bar{j}}/\partial\bar{z}^\beta, \quad \text{etc.},$$

denoting differentiation in $\zeta^i, \bar{\zeta}^j, z^\alpha, \bar{z}^\beta$ by subscripts $i, \bar{j}, \alpha, \bar{\beta}$, respectively. Then on E°, differentiating (2.2) with respect to λ and $\bar{\lambda}$ and then setting $\lambda = 1$, we obtain

$$G(z,\zeta) = \sum_{i,j=1}^{r} G_{i\bar{j}}(z,\zeta)\zeta^i\bar{\zeta}^j. \tag{2.3}$$

Also, differentiating (2.2) with respect to ζ^i and $\bar{\zeta}^j$, we have

$$G_{i\bar{j}}(z,\lambda\zeta) = G_{i\bar{j}}(z,\zeta) \tag{2.4}$$

for all $(z,\zeta) \in E^\circ$ and $\lambda \in \boldsymbol{C}^* = \boldsymbol{C}\setminus\{0\}$, which shows that $G_{i\bar{j}}(z,\zeta)$ is constant along each complex line in the fibres of E. Consequently, if F is smooth on the whole of E, then we have

$$G(z,\zeta) = \sum_{i,j=1}^{r} g_{i\bar{j}}(z)\zeta^i\bar{\zeta}^j,$$

by setting $g_{i\bar{j}}(z) = G_{i\bar{j}}(z, o(z))$, so that $G(z,\zeta)$ corresponds to a Hermitian metric on E provided F is strongly pseudoconvex.

Since $\boldsymbol{C}^*$ acts on E° by scalar multiplication, we obtain the projective bundle $P(E)$ associated with E, which is defined by $P(E) = E^\circ/\boldsymbol{C}^*$ with projection $p : P(E) \to M$. Geometrically, a point $(z,[\zeta])$ in $P(E)$ represents a complex line through ζ in the fibre E_z of E over $z \in M$, and $P(E)$ is a fibre bundle over M with fibre $P_{r-1}(\boldsymbol{C})$, the complex projective space of dimension $r-1$. Let $\tilde{E} = p^{-1}E$ be the pull-back bundle over $P(E)$ induced from E by p. Then we denote by $L(E)$ the tautological line subbundle of $\tilde{E}$, which is defined as follows. Given a point $(z, [\zeta]) \in P(E)$, we take a complex line through ζ in E_z to be a fibre of $L(E)$ over $(z, [\zeta])$. Thus we may summarize the construction in the following diagram:

$$\begin{array}{ccccc} L(E) & \subset & \tilde{E} & \xrightarrow{\tilde{p}} & E \\ & & \downarrow{\scriptstyle \tilde{\pi}} & & \downarrow{\scriptstyle \pi} \\ & & P(E) & \xrightarrow{p} & M \end{array} \tag{2.5}$$

Let $L(E)^{\circ}$ denote $L(E)$ minus its zero section. Then the natural bundle map $\tilde{p} : L(E) \to E$ maps $L(E)^{\circ}$ biholomorphically to E° and collapses the zero section of $L(E)$ to the zero section of E. Thus $L(E)$ may be considered as a blow-up of E along the zero section of E. By means of the biholomorphic correspondence between $L(E)^{\circ}$ and E°, *we have a natural one-to-one correspondence between the Hermitian metrics h on $L(E)$ and the complex Finsler metrics F on E*. Indeed, given a Hermitian metric h on $L(E)$, a complex Finsler metric F on E is defined by setting

$$F^2(z, \zeta) = h(\tilde{p}^{-1}(z, \zeta), \tilde{p}^{-1}(z, \zeta)) \tag{2.6}$$

for $(z, \zeta) \in E^{\circ}$, and vice versa. Note that F is not necessarily smooth along the zero section of E, but h is smooth everywhere on $L(E)$, since it defines an inner product on each fibre of $L(E)$. This explains how naturally complex Finsler metrics emerge in the study of holomorphic vector bundles.

3. Negative vector bundles and complex Finsler metrics

Let E be a holomorphic vector bundle of rank r over a compact complex manifold M of dimension m and $L(E)$ be the tautological line bundle over $P(E)$. Then we say that E is *negative* if the first Chern class $c_1(L(E))$ of $L(E)$ is negative, that is, $c_1(L(E)) < 0$, which means that the line bundle $L(E)$ admits a Hermitian metric h of negative curvature in the following sense. Writing $\hat{h}(\xi)$ for $h(\xi, \xi)$, we consider $\hat{h}$ as a function on a principal $\boldsymbol{C}^*$ bundle $L(E)^{\circ}$ associated with $L(E)$. Then $\omega = \partial \hat{h}$ defines a connection on $L(E)^{\circ}$ with curvature form $\Phi = \bar{\partial}\partial \hat{h}$, and the negativity of the curvature means that, considered as a form on $P(E)$,

$$\sqrt{-1}\Phi = \sqrt{-1}\bar{\partial}\partial \hat{h}$$

is negative definite everywhere. We say that E is *positive* (or *ample*) if the dual bundle E^* of E is negative, which is equivalent to $c_1(L(E))$ being positive, that is, $c_1(L(E)) > 0$. We often write $E < 0$ (resp. $E > 0$) if E is negative (resp. positive).

Given a Hermitian metric h on the tautological line bundle $L(E)$, let F be the complex Finsler metric induced on E from h by (2.6). Following Kobayashi [12], we can determine the explicit relation between the curvature of h and that of

F in the following way. Let $(z,\zeta) = (z^1,\dots,z^m,\zeta^1,\dots,\zeta^r)$ be a local coordinate system for E. This may be considered also as a local coordinate system for $P(E)$ by understanding that ζ^i defines a homogeneous coordinate system $(\zeta^1 : \cdots : \zeta^r)$ for fibres. Setting $Z^i = \zeta^i \circ \tilde{p}$, we take $(z,[\zeta],Z) = (z^1,\dots,z^m,\zeta^1 : \cdots : \zeta^r, Z^1,\dots,Z^r)$ as a local coordinate system for $\tilde{E} = p^{-1}E$. Then the tautological line bundle $L(E)$ over $P(E)$ is defined as a line subbundle of $\tilde{E}$ by the relation $(Z^1 : \cdots : Z^r) = (\zeta^1 : \cdots : \zeta^r)$.

With these conventions understood, we write

$$\Phi = \sum K_{\alpha\bar{\beta}} dz^\alpha \wedge d\bar{z}^\beta + \sum K_{\alpha\bar{j}} dz^\alpha \wedge d\bar{\zeta}^j + \sum K_{i\bar{\beta}} d\zeta^i \wedge d\bar{z}^\beta + \sum K_{i\bar{j}} d\zeta^i \wedge d\bar{\zeta}^j$$

as a $(1,1)$-form on $P(E)$. Remark that, identifying $\tilde{p}^{-1}(z,\zeta) \in L(E)^\circ$ with $(z,\zeta) \in E^\circ$, we obtain from (2.3) and (2.6) that

$$\hat{h}(z,\zeta) = \sum G_{i\bar{j}}(z,\zeta)\zeta^i\bar{\zeta}^j.$$

Hence, utilizing the identity

$$\begin{aligned} &\sum G_{i\bar{j}}(z,\zeta)\zeta^i = G_{\bar{j}}(z,\zeta), \quad \sum G_{i\bar{j}}(z,\zeta)\bar{\zeta}^j = G_i(z,\zeta), \\ &\sum G_{i\bar{j}k}(z,\zeta)\zeta^k = 0, \quad \sum G_{i\bar{j}\bar{l}}(z,\zeta)\bar{\zeta}^l = 0 \end{aligned} \tag{3.1}$$

derived from (2.4), we obtain the following expression, represented in a matrix form, of the curvature of $\hat{h}$:

$$\begin{pmatrix} K_{\alpha\bar{\beta}} & K_{\alpha\bar{j}} \\ K_{i\bar{\beta}} & K_{i\bar{j}} \end{pmatrix} = -\frac{1}{G}\begin{pmatrix} G_{\alpha\bar{\beta}} & G_{\alpha\bar{j}} \\ G_{i\bar{\beta}} & G_{i\bar{j}} \end{pmatrix} + \frac{1}{G^2}\begin{pmatrix} G_\alpha \\ G_i \end{pmatrix}\begin{pmatrix} G_{\bar{\beta}} & G_{\bar{j}} \end{pmatrix}. \tag{3.2}$$

For instance, we have

$$K_{i\bar{j}} = -\frac{1}{G}G_{i\bar{j}} + \frac{1}{G^2}\sum G_{i\bar{l}}\bar{\zeta}^l G_{k\bar{j}}\zeta^k,$$

so that for $(z,[\zeta],Z) \in \tilde{E}$

$$\begin{aligned} \sum K_{i\bar{j}} Z^i \bar{Z}^j = -\frac{1}{G^2}\Big\{ &\Big(\sum G_{i\bar{j}} Z^i \bar{Z}^j\Big)\Big(\sum G_{k\bar{l}}\zeta^k\bar{\zeta}^l\Big) \\ &- \Big(\sum G_{i\bar{l}} Z^i \bar{\zeta}^l\Big)\Big(\sum G_{k\bar{j}}\zeta^k \bar{Z}^j\Big)\Big\}. \end{aligned}$$

This shows that $\sum G_{i\bar{j}} Z^i \bar{Z}^j > 0$ if and only if $\sum K_{i\bar{j}} Z^i \bar{Z}^j < 0$ except when $Z^i = c\zeta^i$ for $i = 1,\dots,r$. Indeed, since the curvature form Φ is actually a form on $P(E)$, it degenerates along the fibres of $L(E)$ and the exceptional direction $(Z^1 : \cdots : Z^r)$ above is precisely the fibre direction of $L(E)$. Consequently, we see that $\sqrt{-1}\Phi$, *restricted to each fibre of* $P(E) \to M$, *is negative definite if and only if* $(G_{i\bar{j}})$ *is positive definite, that is,* F *is strongly pseudoconvex.*

Given a complex Finsler metric F on E, recall that $G_{i\bar{j}}(z,\zeta)$ is constant along each complex line in the fibres of E and hence gives rise to a function $G_{i\bar{j}}(z,[\zeta])$

defined on $P(E)$. In order to see other relations of the curvature Φ of h to that of F, we now set

$$\tilde{G}(z,[\zeta],Z) = \sum_{i,j=1}^{r} G_{i\bar{j}}(z,[\zeta])Z^i\bar{Z}^j \tag{3.3}$$

for $(z,[\zeta],Z) \in \tilde{E}$. If F is strongly pseudoconvex, this defines a Hermitian metric on the pull-back bundle $\tilde{E}$ over $P(E)$, which coincides with $G(z,\zeta)$ when restricted to $L(E)^{\circ}$. Thus we can apply results from Hermitian vector bundles to $(\tilde{E},\tilde{G})$.

From now on, we assume that F is strongly pseudoconvex. Let $(G^{\bar{k}l})$ be the inverse matrix of $(G_{i\bar{j}})$. Then the connection form $\omega = (\omega^i_j)$ of the Hermitian connection D of the Hermitian vector bundle $(\tilde{E},\tilde{G})$ is given by

$$\omega^i_j = \sum G^{\bar{k}i}\partial G_{j\bar{k}} = \sum \Gamma^i_{j\alpha}dz^\alpha + \sum \Gamma^i_{jk}d\zeta^k, \tag{3.4}$$

and the curvature form $\Omega = (\Omega^i_j)$ of its curvature $R = D\circ D$ is given by $\Omega^i_j = \bar{\partial}\omega^i_j$, which we can write

$$\Omega^i_j = \sum R^i_{j\alpha\bar{\beta}}dz^\alpha\wedge d\bar{z}^\beta + \sum R^i_{j\alpha\bar{l}}dz^\alpha\wedge d\bar{\zeta}^l + \sum R^i_{jk\bar{\beta}}d\zeta^k\wedge d\bar{z}^\beta + \sum R^i_{jk\bar{l}}d\zeta^k\wedge d\bar{\zeta}^l.$$

Setting $R_{i\bar{j}\alpha\bar{\beta}} = \sum G_{k\bar{j}}R^k_{i\alpha\bar{\beta}}$, etc., and utilizing (2.3) and (3.1) together with the identity

$$\sum G_{ik}(z,\zeta)\zeta^i = 0, \quad \sum G_{\bar{j}\bar{l}}(z,\zeta)\bar{\zeta}^j = 0$$

derived from (2.4), we then obtain

$$\begin{aligned} &\sum R_{i\bar{j}\alpha\bar{\beta}}\zeta^i\bar{\zeta}^j = -G_{\alpha\bar{\beta}} + \sum G^{\bar{l}k}G_{k\bar{\beta}}G_{\bar{l}\alpha}, \quad \sum R_{i\bar{j}\alpha\bar{l}}\bar{\zeta}^j = \sum R_{i\bar{j}\alpha\bar{l}}\bar{\zeta}^l = 0,\\ &\sum R_{i\bar{j}k\bar{\beta}}\zeta^i = \sum R_{i\bar{j}k\bar{\beta}}\zeta^k = 0, \quad \sum R_{i\bar{j}k\bar{l}}\zeta^i = \sum R_{i\bar{j}k\bar{l}}\zeta^k = 0. \end{aligned} \tag{3.5}$$

To each $(z,[\zeta],Z) \in \tilde{E}$, we associate the following Hermitian form on $P(E)$:

$$\begin{aligned} \Psi(z,[\zeta],Z) = &\sum R_{i\bar{j}\alpha\bar{\beta}}Z^i\bar{Z}^j dz^\alpha d\bar{z}^\beta + \sum R_{i\bar{j}\alpha\bar{l}}Z^i\bar{Z}^j dz^\alpha d\bar{\zeta}^l\\ &+ \sum R_{i\bar{j}k\bar{\beta}}Z^i\bar{Z}^j d\zeta^k d\bar{z}^\beta + \sum R_{i\bar{j}k\bar{l}}Z^i\bar{Z}^j d\zeta^k d\bar{\zeta}^l. \end{aligned}$$

Restricting this to $L(E) \subset \tilde{E}$ and using (3.5), we have

$$\Psi(z,[\zeta],\zeta) = \sum R_{i\bar{j}\alpha\bar{\beta}}\zeta^i\bar{\zeta}^j dz^\alpha d\bar{z}^\beta = \sum\left(-G_{\alpha\bar{\beta}} + \sum G^{k\bar{l}}G_{k\bar{\beta}}G_{\bar{l}\alpha}\right)dz^\alpha d\bar{z}^\beta.$$

We say that F has *negative* (resp. *positive*) *curvature* if the quantity $(\sum R_{i\bar{j}\alpha\bar{\beta}}\zeta^i\bar{\zeta}^j)$ is negative (resp. positive) for all $(z,\zeta) \in E^{\circ}$, that is, $\Psi(z,[\zeta],\zeta)$ is negative (resp. positive) for all $(z,[\zeta],\zeta) \in L(E)^{\circ}$.

Now, in order to see the signs of these, given a point (z_0,ζ_0) in E, we take a normal coordinate system $(z^1,\dots,z^m,\zeta^1,\dots,\zeta^r)$ at (z_0,ζ_0), which satisfies

$$G_{i\bar{j}}(z_0,\zeta_0) = \delta_{ij}, \quad G_{i\bar{j}\alpha}(z_0,\zeta_0) = G_{i\bar{j}\bar{\beta}}(z_0,\zeta_0) = 0.$$

With respect to this coordinate system, differentiating (2.3) by z^α and $\bar{z}^\beta$, we have

$$G_\alpha(z_0,\zeta_0) = G_{\bar{\beta}}(z_0,\zeta_0) = 0,$$

and, differentiating (3.1) by z^α and $\bar{z}^\beta$,

$$G_{i\alpha}(z_0, \zeta_0) = G_{i\bar{\beta}}(z_0, \zeta_0) = G_{\bar{j}\alpha}(z_0, \zeta_0) = G_{\bar{j}\bar{\beta}}(z_0, \zeta_0) = 0.$$

Hence, noting (3.2), we find that

$$\sum R_{i\bar{j}\alpha\bar{\beta}}(z_0, \zeta_0)\zeta_0^i \bar{\zeta}_0^j = -G_{\alpha\bar{\beta}}(z_0, \zeta_0) = G(z_0, \zeta_0) K_{\alpha\bar{\beta}}(z_0, \zeta_0),$$

$$\Psi(z_0, [\zeta_0], \zeta_0) = G(z_0, \zeta_0) \sum K_{\alpha\bar{\beta}}(z_0, \zeta_0) dz^\alpha d\bar{z}^\beta.$$

From these observations and formulae we can now conclude the following

Theorem 1 (Kobayashi [12, 13]**).** *Let E be a holomorphic vector bundle over a complex manifold M. Let h be a Hermitian metric on the tautological line bundle $L(E)$ over $P(E)$ and F the corresponding complex Finsler metric on E. Then the following hold:*

(1) *F is strongly pseudoconvex if and only if the curvature form $\sqrt{-1}\Phi$ of h is negative definite along each fibre of $P(E) \to M$.*
(2) *$\sqrt{-1}\Phi$ is negative definite on $P(E)$ if and only if F is strongly pseudoconvex and has negative curvature.*

As a result, we obtain the following differential geometric characterization of a negative vector bundle E in terms of complex Finsler metrics on E.

Corollary 1 ([12, 13])**.** *A holomorphic vector bundle E over a compact complex manifold M is negative if and only if it admits a strongly pseudoconvex Finsler metric F with negative curvature.*

Remark 1 ([13])**.** *Let E be a holomorphic vector bundle over M, and E^* the dual vector bundle of E. Let $\tilde{E} = p^{-1}E$ be the pull-back bundle over $P(E)$ as in (2.5), and consider the dual vector bundle $\tilde{E}^*$ of $\tilde{E}$* over *$P(E)$. Recall that, given a strongly pseudoconvex Finsler metric F on E, we obtain a Hermitian metric on $\tilde{E}$ defined by*

$$\tilde{G}(z, [\zeta], Z) = \sum G_{i\bar{j}}(z, [\zeta]) Z^i \bar{Z}^j$$

as in (3.3). Taking the local frame field $(s^1, \ldots, s^r)$ of E^* dual to $(s_1, \ldots, s_r)$ of E, we may introduce the local fibre coordinate system $Z^* = (Z_1, \ldots, Z_r)$ for $\tilde{E}^*$ dual to $(Z^1, \ldots, Z^r)$ of $\tilde{E}$. Then

$$\tilde{G}^*(z, [\zeta], Z^*) = \sum G^{i\bar{j}}(z, [\zeta]) Z_i \bar{Z}_j$$

defines a Hermitian metric on $\tilde{E}^*$.

From (3.4) we see that the connection form $({}^*\omega^i_j)$ for $(\tilde{E}^*, \tilde{G}^*)$ is given by

$${}^*\omega^i_j = \sum G_{j\bar{k}} \partial G^{\bar{k}i} = -\omega^i_j.$$

It then follows that the curvature of $\tilde{G}$ is negative if and only if the curvature of $\tilde{G}^*$ is positive. Hence E^* is positive or ample if and only if $\tilde{E}^*$ admits a strongly pseudoconvex Finsler metric with positive curvature.

4. Frankel conjecture versus Hartshorne conjecture

Around 1979 the following Hartshorne conjecture in algebraic geometry, over the field of complex numbers, and the Frankel conjecture in complex differential geometry were proved by Mori [19] and by Siu-Yau [24], independently.

Theorem 2 (Hartshorne conjecture [10, 19]**).** *Let M be a compact complex manifold of dimension m with ample holomorphic tangent bundle. Then M is biholomorphic to a complex projective space $P_m(\boldsymbol{C})$ of dimension m.*

Theorem 3 (Frankel conjecture [7, 16, 24]**).** *Let M be a compact Kähler manifold of dimension m with positive holomorphic bisectional curvature. Then M is biholomorphic to a complex projective space $P_m(\boldsymbol{C})$ of dimension m.*

Note that these theorems are obvious if $m = 1$ by the Gauss-Bonnet theorem. Prior to the complete resolution by Mori and Siu-Yau, Theorem 2 was proved by Hartshorne [10] in dimension 2, whereas Theorem 3 was proved by Andreotti-Frankel [7] in dimension 2, and then in dimension 3 by Mabuchi [16] using the result of Kobayashi-Ochiai [14]. Also, it should be remarked that, since the positivity of the holomorphic bisectional curvature of a Kähler metric, or more generally a Hermitian metric, implies the positivity of the holomorphic tangent bundle, Theorem 2 proves Theorem 3 as its corollary.

In both proofs by Mori and Siu-Yau, the main point of the argument is to show that the M in question contains a specific rational curve. Mori [19] achieved this by using the method of algebraic geometry of positive characteristic, and therefore his proof is purely algebraic in nature. On the other hand, Siu-Yau used harmonic maps and the characterization of complex projective spaces due to Kobayashi-Ochiai [15], thus employing the method of Kähler geometry to answer a question in Kähler geometry. We now sketch the barest outline of the proof of the Frankel conjecture by Siu-Yau [24].

According to the result of Kobayashi-Ochiai, a compact complex manifold of dimension m is biholomorphic to a complex projective space $P_m(\boldsymbol{C})$ of dimension m if and only if the first Chern class $c_1(M)$ of M satisfies an inequality

$$c_1(M) \geq (m+1)c_1(L)$$

for some positive holomorphic line bundle L over M. To put this to use, we recall that by a result of Bishop-Goldberg [4, 8], if M has positive holomorphic bisectional curvature, then $\dim H^2(M, \boldsymbol{R}) = 1$, where $H^2(M, \boldsymbol{R})$ denotes the second cohomology group of M with real coefficients. Since M then has positive Ricci curvature, a theorem of Kobayashi [11] implies that M is simply connected. Then it follows from $H_1(M, \boldsymbol{Z}) = \{0\}$ and $H^2(M, \boldsymbol{R}) = \boldsymbol{R}$ that $H^2(M, \boldsymbol{Z}) = \boldsymbol{Z}$ and $H_2(M, \boldsymbol{Z}) = \boldsymbol{Z}$ modulo torsions. Let ω be the generator of $H^2(M, \boldsymbol{Z})$ which is a positive multiple of the Kähler form of M. Then ω is positive and by Hodge theory, there is a positive holomorphic line bundle L with $c_1(L) = \omega$. Hence by the result of Kobayashi-Ochiai, it suffices to show that $c_1(M) = \lambda\omega$ for some integer

$\lambda \geq m+1$. If σ denotes the generator of the free part of $H_2(M, \boldsymbol{Z})$ such that $\omega(\sigma)=1$, then it suffices in turn to show that

$$c_1(M)(\sigma) \geq m+1. \tag{4.1}$$

This can be done in the following way. Since M is simply connected, by the Hurewitz isomorphism theorem, $H_2(M, \boldsymbol{Z}) \cong \pi_2(M)$, where $\pi_2(M)$ denotes the second homotopy group of M. Therefore we may identify σ with a smooth map $f: S^2 \to M$ such that $[f]=\sigma$. Then (4.1) is accomplished if we show that f is homotopic to a rational curve in M, that is, a holomorphic map from $P_1(\boldsymbol{C}) \cong S^2$ to M. Indeed, let T_M and T_P denote the holomorphic tangent bundles of M and $P_1(\boldsymbol{C})$, respectively, and suppose f is an immersion, for simplicity. Then, a well-known theorem due to Grothendieck [9] states that the normal bundle $f^{-1}T_M/T_P$ splits into a direct sum of holomorphic line bundles L_i, so that $f^{-1}T_M \cong T_P \oplus L_1 \cdots \oplus L_{m-1}$. Hence

$$c_1(f^{-1}T_M) = c_1(T_P) + c_1(L_1) + \cdots + c_1(L_{m-1}).$$

On the other hand, by the curvature condition, the holomorphic tangent bundle of M is positive, and hence each L_i is also positive, so that $c_1(L_i)([P_1(\boldsymbol{C})]) \geq 1$. Hence we obtain (4.1), since $c_1(T_P)([P_1(\boldsymbol{C})]) = 2$.

The existence of such a rational curve is obtained in the following manner. By a theorem of Sacks-Uhlenbeck [23], the infimum of the energies of smooth maps from S^2 to M representing the generator of $\pi_2(M)$ can be achieved by a sum of energy-minimizing harmonic maps $f_j : S^2 \to M$, where $1 \leq j \leq s$. Then the key step of the proof is to show that each f_j is either holomorphic or antiholomorphic. To this end we need to look at the second variation formula of the energy functional for a smooth variation, parametrized by a complex number, of smooth maps from S^2 to M. Then, for an energy-minimizing harmonic map $f_j : S^2 \to M$, the second variation of the energy functional must be nonnegative. However, by a direct computation, we find that if the holomorphic bisectional curvature of M is positive, this second variation turns out to be nonpositive, so we conclude that f_j is either holomorphic or antiholomorphic. Hence the proof is finished if we show that $s=1$. This can be done as follows. In case $s>1$, we can holomorphically deform the images of some holomorphic f_j and antiholomorphic f_k so that they are tangential to each other at some point. By removing a small disk centered at the point of contact from each and joining the two disk boundaries by a suitable surface, we obtain a smooth map from S^2 to M with energy smaller than the minimum energy. Thus $s=1$ and the image of f_1 is a rational curve representing a generator of the free part of $H_2(M, \boldsymbol{Z})$.

As remarked above, since Theorem 3 follows from Theorem 2 as a special case, this differential geometric proof of the Frankel conjecture by Siu-Yau, in the framework of Kähler geometry, does not cover the Hartshorne conjecture in its general form. However, since the Hartshorne conjecture in the case of the complex number field is a question within complex differential geometry, it is of great interest to find an independent proof of Theorem 2 in the framework of complex differential geometry. Concerning this problem, we now recall a theorem of Kobayashi, Corollary 1 in Section 3, which states that a holomorphic vector bundle over a compact complex manifold is negative if and only if it admits a strongly pseudoconvex Finsler metric with negative curvature. In view of the above method employed by Siu-Yau, one might speculate if the harmonic map theory has its counterpart in complex Finsler geometry, which we discuss in the next section.

5. Harmonic maps into complex Finsler manifolds

Let M be a complex manifold of dimension m. We denote the real tangent bundle of M by TM and the holomorphic tangent bundle of M by $\pi : T^{1,0}M \to M$, respectively. Note that $T^{1,0}M$ is a holomorphic vector bundle of rank m over M. Indeed, for a local coordinate system $z = (z^1, \dots, z^m)$ for M, a holomorphic tangent vector v of M is written as

$$v = \sum_{\alpha=1}^{m} v^\alpha \frac{\partial}{\partial z^\alpha},$$

and we may take $(z, v) = (z^1, \dots, z^m, v^1, \dots, v^m)$ as a local coordinate system for $T^{1,0}M$. Let $F : T^{1,0}M \to \boldsymbol{R}$ be a strongly pseudoconvex Finsler metric defined on $T^{1,0}M$, which we call from now on a strongly pseudoconvex Finsler metric on M for short. Also, we set $\tilde{M} = T^{1,0}M \setminus o(M)$, that is, $\tilde{M}$ denotes $T^{1,0}M$ minus its zero section.[1] Then, as in Section 2, $G = F^2$ is expressed on $\tilde{M}$ as

$$G(z,v) = \sum_{\mu,\nu=1}^{m} G_{\mu\bar{\nu}}(z,v) v^\mu \bar{v}^\nu, \quad G_{\mu\bar{\nu}} = \partial^2 G / \partial v^\mu \partial \bar{v}^\nu.$$

Now, let N be a compact Riemann surface, that is, a compact complex manifold of dimension 1, and $f : N \to M$ a smooth map from N to M. We denote a local coordinate system for N by w, and express f locally as

$$f(w) = \left(f^1(w), \dots, f^m(w)\right) = \left(f^\alpha(w)\right).$$

Note that the differential $df : TN \to TM$ of f extends to a complex linear map between the complexified tangent bundles $T_{\boldsymbol{C}}N$ and $T_{\boldsymbol{C}}M$. According to the decompositions

$$T_{\boldsymbol{C}}N = T^{1,0}N \oplus T^{0,1}N, \quad T_{\boldsymbol{C}}M = T^{1,0}M \oplus T^{0,1}M \tag{5.1}$$

into the holomorphic tangent bundles, $T^{1,0}N$ and $T^{1,0}M$, and the antiholomorphic tangent bundles, $T^{0,1}N$ and $T^{0,1}M$, of N and M, respectively, we obtain from $df : T_{\boldsymbol{C}}N \to T_{\boldsymbol{C}}M$ the endomorphisms

$$\partial f : T^{1,0}N \to T^{1,0}M, \quad \bar{\partial} f : T^{0,1}N \to T^{1,0}M,$$

[1]Note that in Section 2, we denote $T^{1,0}M \setminus o(M)$ by $(T^{1,0}M)^\circ$, and by $\tilde{M}$ the pull-back $\pi^{-1}(T^{1,0}M)$ of $T^{1,0}M$ over the projective bundle $P(T^{1,0}M)$ associated with $T^{1,0}M$, instead.

which are defined in local coordinates respectively by

$$\partial f\left(\frac{\partial}{\partial w}\right)=\sum\frac{\partial f^\alpha}{\partial w}\frac{\partial}{\partial z^\alpha},\quad \bar\partial f\left(\frac{\partial}{\partial \bar w}\right)=\sum\frac{\partial f^\alpha}{\partial \bar w}\frac{\partial}{\partial z^\alpha}.$$

Then we define the *$\bar\partial$-energy* $E_{\bar\partial}(f)$ of f by

$$E_{\bar\partial}(f)=\int_N F^2\left(f(w),\bar\partial f\left(\frac{\partial}{\partial \bar w}\right)\right)\frac{\sqrt{-1}}{2}dw\wedge d\bar w. \tag{5.2}$$

It should be remarked that, by virtue of the $(1,1)$-homogeneity (2.2) of F^2, the definition (5.2) is independent of the choice of a local coordinate system w for N, and thus the $\bar\partial$-energy of f is well defined.

We now consider a smooth variation of $f=f_0$ through smooth maps

$$f_t:N\to M,\qquad t\in\Delta=\{z\in \boldsymbol{C}\mid |z|<\epsilon\}, \tag{5.3}$$

parametrized by an open disk Δ in $\boldsymbol{C}$. In order to compute the first variation of the $\bar\partial$-energy $E_{\bar\partial}(f_t)$, let $(G^{\bar\tau\sigma})$ denote the inverse matrix of $(G_{\mu\bar\nu})$ and set

$$\Gamma^\mu_{\nu,\alpha}=\sum G^{\bar\tau\mu}\left(\frac{\partial G_{\nu\bar\tau}}{\partial z^\alpha}-\sum G^{\bar\rho\sigma}\frac{\partial G_{\nu\bar\tau}}{\partial v^\sigma}\frac{\partial^2 G}{\partial\bar v^\rho\partial z^\alpha}\right). \tag{5.4}$$

The geometric meaning of $\Gamma^\mu_{\nu,\alpha}$ as components of the canonical connection defined by the Finsler metric F will be explained in the next section. Then we want to compute the first variation

$$\begin{aligned}\frac{\partial}{\partial t}E_{\bar\partial}(f_t)&=\frac{\partial}{\partial t}\int_N F^2\left(f_t(w),\bar\partial f_t\left(\frac{\partial}{\partial\bar w}\right)\right)\frac{\sqrt{-1}}{2}dw\wedge d\bar w\\ &=\int_N\frac{\partial}{\partial t}F^2\left(f_t(w),\bar\partial f_t\left(\frac{\partial}{\partial\bar w}\right)\right)\frac{\sqrt{-1}}{2}dw\wedge d\bar w.\end{aligned} \tag{5.5}$$

Here we note that a technical intricacy arises in computing the integrand on the right-hand side of (5.5), since F is not smooth in general along the zero section of $T^{1,0}M$. (cf. Section 2). However, the $(1,1)$-homogeneity of F^2 implies that F^2 is once differentiable and has Lipschitz continuous derivatives everywhere on $T^{1,0}M$, and hence the chain rule for distribution derivatives holds in this context. This assures that it suffices to differentiate the integrand as usual outside the zero section of $T^{1,0}M$, that is, whenever $\bar\partial f_t(\partial/\partial\bar w)\neq 0$, and understand the result vanishing when $\bar\partial f_t(\partial/\partial\bar w)=0$ (See for instance [17]).

With this understood, by a straightforward computation, we then obtain the following formula for the first variation:

$$\begin{aligned}\frac{\partial}{\partial t}E_{\bar\partial}(f_t)=&-\int_N G_{\mu\bar\nu}Q^\mu\left(\overline{\frac{\partial f_t^\nu}{\partial \bar t}}\right)\frac{\sqrt{-1}}{2}dw\wedge d\bar w\\ &-\int_N G_{\mu\bar\nu}\left(\frac{\partial f_t^\mu}{\partial t}\right)\overline{Q^\nu}\frac{\sqrt{-1}}{2}dw\wedge d\bar w,\end{aligned} \tag{5.6}$$

where $Q^\mu(f_t(w), \bar\partial f_t(\partial/\partial\bar w))$ is given by

$$Q^\mu = \frac{\partial^2 f_t^\mu}{\partial w \partial \bar w} + \sum \Gamma^\mu_{\sigma,\rho} \frac{\partial f_t^\rho}{\partial w}\frac{\partial f_t^\sigma}{\partial \bar w} - \sum G^{\bar\tau\mu}\frac{\partial G}{\partial \bar v^\rho}\left(\Gamma^{\bar\rho}_{\bar\tau,\bar\sigma} - \Gamma^{\bar\rho}_{\bar\sigma,\bar\tau}\right)\overline{\frac{\partial f_t^\sigma}{\partial \bar w}}, \tag{5.7}$$

which is understood to be zero when $\bar\partial f_t(\partial/\partial\bar w) = 0$.

Now we recall that a strongly pseudoconvex Finsler metric F is said to be *weakly Kähler* if it satisfies for all $(z,v) \in \tilde M$

$$\sum \frac{\partial G}{\partial v^\rho}\left(\Gamma^\rho_{\tau,\sigma} - \Gamma^\rho_{\sigma,\tau}\right) v^\sigma = 0. \tag{5.8}$$

Geometrically, this condition means that a suitable kind of contraction of the torsion of the canonical connection defined by F vanishes identically, as will be explained in the next section.

In the following, we assume that $F : T^{1,0}M \to \boldsymbol{R}$ is a weakly Kähler Finsler metric on M. Then it follows from (5.6) and (5.7) that $f = f_0$ is a critical point of the $\bar\partial$-energy functional if and only if it satisfies the equations

$$\frac{\partial^2 f^\mu}{\partial w \partial \bar w}(w) + \sum_{\rho,\sigma=1}^{m} \Gamma^\mu_{\sigma,\rho}\left(f(w), \bar\partial f\left(\frac{\partial}{\partial \bar w}\right)\right)\frac{\partial f^\rho}{\partial w}(w)\frac{\partial f^\sigma}{\partial \bar w}(w) = 0 \tag{5.9}$$

for all $w \in N$ and $1 \le \mu \le m$, where $\Gamma^\mu_{\sigma,\rho}$ are defined by (5.4) on $\tilde M$ and interpreted as zero whenever $\bar\partial f(\partial/\partial\bar w) = 0$. In fact, by the $(1,1)$-homogeneity of G again, it is verified that the second term on the left-hand side of (5.9) reduces to

$$\sum G^{\bar\tau\mu}\frac{\partial G_{\sigma\bar\tau}}{\partial z^\rho}\frac{\partial f^\rho}{\partial w}\frac{\partial f^\sigma}{\partial \bar w}.$$

To sum up, we arrive at the following

Definition 1. *Let M be a complex manifold equipped with a weakly Kähler Finsler metric $F : T^{1,0}M \to \boldsymbol{R}$. Then a smooth map $f : N \to M$ from a compact Riemann surface N to M is called a* harmonic map *if it satisfies the equations* (5.9) *on N.*

From this definition, we immediately see that a holomorphic map or an antiholomorphic map $f : N \to M$ is a harmonic map. Also, we remark that if in particular M is a Kähler manifold and F is the corresponding complex Finsler metric as in (2.1), then F is obviously a weakly Kähler Finsler metric and the equations (5.9) reduce to the well-known harmonic map equation for maps from a Riemann surface N to a Kähler manifold M, since $\Gamma^\mu_{\sigma,\rho}$ coincide with the Christoffel symbols of the Levi-Civita connection of M (cf. [6]).

Remark 2. *Our definition of the $\bar\partial$-energy* (5.2) *deals only with the case when the source manifold N is a Riemann surface. For other sorts of definitions of the energy functional for maps between Finsler manifolds or from Finsler manifolds into Riemannian manifolds, see* [5] *and* [18].

6. Chern-Finsler connection

We are now concerned with the second variation formula of the $\bar{\partial}$-energy functional. In computing this, as inferred from the formula for the usual energy functional (cf. [6]), it is convenient to utilize the canonical connection associated with a strongly pseudoconvex Finsler metric. This being the case, we summarize in this section a brief overview of the theory of connection in complex Finsler geometry. For more details the reader is referred to a monograph by Abate-Patrizio [1]. We mostly follow the notation used in this monograph.

Let M be a complex manifold of dimension m, and $\pi : T^{1,0}M \to M$ the holomorphic tangent bundle of M. As in Section 5, $\tilde{M} = T^{1,0}M \setminus o(M)$ will denote $T^{1,0}M$ minus the zero section. We denote local coordinate systems for M and $T^{1,0}M$ by $z = (z^1, \dots, z^m)$ and $(z,v) = (z^1, \dots, z^m, v^1, \dots, v^m)$, respectively, and set for simplicity

$$\partial_\alpha = \frac{\partial}{\partial z^\alpha}, \quad \dot{\partial}_\mu = \frac{\partial}{\partial v^\mu}, \qquad 1 \le \alpha, \mu \le m,$$

which give a local frame field of the holomorphic tangent bundle $T^{1,0}\tilde{M}$ of $\tilde{M}$. Also, we denote $(z,v) \in T^{1,0}M$ simply by $v \in T^{1,0}M$, when there is no fear of confusion.

Let $F : T^{1,0}M \to \boldsymbol{R}$ be a strongly pseudoconvex Finsler metric defined on $T^{1,0}M$, and set $G = F^2$. In order to distinguish the derivatives with respect to the z-coordinates from those with respect to the v-coordinates, we indicate, as in the previous section, the differentiation with respect to z-coordinates by indices after a comma. Thus we write, for instance,

$$\begin{aligned} G_\mu &= \partial G/\partial v^\mu, & G_{\mu\bar{\nu}} &= \partial^2 G/\partial v^\mu \partial \bar{v}^\nu, & G_{\mu,\alpha} &= \partial^2 G/\partial v^\mu \partial z^\alpha, \\ G_{,\alpha} &= \partial G/\partial z^\alpha, & G_{,\alpha\bar{\beta}} &= \partial^2 G/\partial z^\alpha \partial \bar{z}^\beta, & G_{\mu\bar{\nu},\bar{\alpha}} &= \partial G_{\mu\bar{\nu}}/\partial \bar{z}^\alpha. \end{aligned}$$

Recall that, as in (2.3), G is expressed on $\tilde{M}$ as

$$G(z,v) = \sum G_{\mu\bar{\nu}}(z,v) v^\mu \bar{v}^\nu. \tag{6.1}$$

In Section 3, we associated with (6.1) a Hermitian metric

$$\tilde{G}(z,[v],Z) = \sum G_{\mu\bar{\nu}}(z,[v]) Z^\mu \bar{Z}^\nu$$

defined on the pull-back $\pi^{-1}(T^{1,0}M)$ of $T^{1,0}M$ over the projective bundle associated with $T^{1,0}M$, or more primitively over $\tilde{M}$, and applied results from Hermitian vector bundles to $(\pi^{-1}(T^{1,0}M), \tilde{G})$.

However, we do not take this point of view in the following, but instead will identify the pull-back bundle $\pi^{-1}(T^{1,0}M)$ with the complex vertical subbundle $\mathcal{V}$ of the holomorphic tangent bundle $T^{1,0}\tilde{M}$ of $\tilde{M}$, that is, with a subbundle living in the second order tangent bundle of M. In other words, we replace the given strongly pseudoconvex Finsler metric on $T^{1,0}M$ with a Hermitian metric on a suitable subbundle of the holomorphic tangent bundle of $\tilde{M}$, by going one step upstairs, as summarized in the following diagram:

$$
\begin{array}{ccccc}
 & & T^{1,0}\tilde{M} & = & \mathcal{V} \oplus \mathcal{H} \\
 & & \downarrow{\scriptstyle \tilde{\pi}} & & \\
\pi^{-1}(T^{1,0}M) & \longrightarrow & \tilde{M} & \subset & T^{1,0}M \\
\downarrow & & \downarrow{\scriptstyle \pi} & & \\
\tilde{M} & \xrightarrow{\ \pi\ } & M & &
\end{array}
\tag{6.2}
$$

To be more precise, we proceed as follows. Let $\tilde{\pi} : T^{1,0}\tilde{M} \to \tilde{M}$ denote the holomorphic tangent bundle of $\tilde{M}$. Then the differential $d\pi : T_{\mathbf{C}}\tilde{M} \to T_{\mathbf{C}}M$ of $\pi : \tilde{M} \to M$ defines the *vertical bundle* $\mathcal{V}$ over $\tilde{M}$ by

$$\mathcal{V} = \operatorname{Ker} d\pi \cap T^{1,0}\tilde{M} \subset T^{1,0}\tilde{M},$$

which yields a holomorphic vector bundle of rank m over $\tilde{M}$. A local frame field of $\mathcal{V}$ is given by $\{\dot{\partial}_1, \ldots, \dot{\partial}_m\}$, and a natural section $\iota : \tilde{M} \to \mathcal{V}$, called the *radial vertical field*, is well defined for $(z,v) \in \tilde{M}$ by

$$\iota(v) = \iota\left(\sum v^\alpha \left(\frac{\partial}{\partial z^\alpha}\right)_z\right) = \sum v^\mu \left(\dot{\partial}_\mu\right)_v .$$

Associated with (6.1), we now define a Hermitian metric on the vertical bundle $\mathcal{V}$ by

$$\langle Z_1, Z_2 \rangle_v = \sum_{\mu,\nu=1}^{m} G_{\mu\bar{\nu}}(z,v) Z_1^\mu \bar{Z}_2^\nu, \tag{6.3}$$

where $(z,v) \in \tilde{M}$ and $Z_1, Z_2 \in \mathcal{V}_v = \mathcal{V} \cap \tilde{\pi}^{-1}(z,v)$. Note that the definition of the Hermitian metric (6.3) is well defined, and the radial vertical field $\iota : \tilde{M} \to \mathcal{V}$ yields an isometric embedding of $\tilde{M}$ into $\mathcal{V}$, since

$$G(z,v) = \sum G_{\mu\bar{\nu}}(z,v) v^\mu \bar{v}^\nu = \langle \iota(v), \iota(v) \rangle.$$

Let $D : \Gamma(\mathcal{V}) \to \Gamma(T^*_{\mathbf{C}}\tilde{M} \otimes \mathcal{V})$ be the Hermitian connection of the Hermitian vector bundle $(\mathcal{V}, \langle\ ,\ \rangle)$, where $\Gamma(\,\cdot\,)$ denotes the space of smooth sections. Let ∇ denote the covariant differentiation defined by D, and define a bundle map $\Lambda : T^{1,0}\tilde{M} \to \mathcal{V}$ by $\Lambda(X) = \nabla_X \iota$. The *horizontal bundle* $\mathcal{H}$ over $\tilde{M}$ is then defined by $\mathcal{H} = \operatorname{Ker} \Lambda$, which is the subbundle of $T^{1,0}\tilde{M}$ consisting of vectors with respect to which ι is parallel. Then it is verified that $T^{1,0}\tilde{M} = \mathcal{V} \oplus \mathcal{H}$ and a natural local frame field $\{\delta_1, \ldots, \delta_m\}$ of $\mathcal{H}$ is given by

$$\delta_\alpha = \partial_\alpha - \sum \Gamma^\mu_{,\alpha} \dot{\partial}_\mu, \quad \Gamma^\mu_{,\alpha} = \sum G^{\bar{\tau}\mu} G_{\bar{\tau},\alpha}, \tag{6.4}$$

where $(G^{\bar{\tau}\sigma})$ is the inverse matrix of $(G_{\mu\bar{\nu}})$ as in the previous section.

Associated with the decomposition $T^{1,0}\tilde{M} = \mathcal{V} \oplus \mathcal{H}$, we have the *horizontal map* $\Theta : \mathcal{V} \to \mathcal{H}$ given locally by $\Theta(\dot{\partial}_\mu) = \delta_\mu$ for $1 \leq \mu \leq m$, and a natural section

$$\chi = \Theta \circ \iota : \tilde{M} \to \mathcal{H},$$

called the *radial horizontal field*, such that

$$\chi\left(\sum v^\alpha \frac{\partial}{\partial z^\alpha}\right) = \sum v^\alpha \delta_\alpha.$$

More generally, for each $(z, v) \in \tilde{M}$ we have a map $\chi_v : T^{1,0}_z M \to \mathcal{H}_v$ given locally by

$$\chi_v\left(\sum w^\alpha \left(\frac{\partial}{\partial z^\alpha}\right)_z\right) = \sum w^\alpha \left(\delta_\alpha\right)_v.$$

This allows us to canonically lift, in (6.2), a vector field on M to a horizontal vector field on $\tilde{M}$. Indeed, given a smooth vector field $\xi \in \Gamma(\tilde{M})$, the *horizontal lift* ξ^H of ξ is defined by

$$\xi^H(v) = \chi_v\left(\xi(z)\right)$$

for $(z, v) \in \tilde{M}$.

Using the horizontal map $\Theta : \mathcal{V} \to \mathcal{H}$, we can transfer the Hermitian metric $\langle\ ,\ \rangle$ on $\mathcal{V}$, defined by (6.3), to $\mathcal{H}$ by setting

$$\langle H_1, H_2\rangle_v = \langle \Theta^{-1}(H_1), \Theta^{-1}(H_2)\rangle_v,$$

where $(z, v) \in \tilde{M}$ and $H_1, H_2 \in \mathcal{H}_v = \mathcal{H} \cap \tilde{\pi}^{-1}(z, v)$. Then a Hermitian metric $\langle\ ,\ \rangle$ on $T^{1,0}\tilde{M}$ canonically associated with F is defined by requiring $\mathcal{H}$ to be orthogonal to $\mathcal{V}$, so that $\Theta : \mathcal{V} \to \mathcal{H}$ and $\chi : \tilde{M} \to \mathcal{H}$ are isometric embeddings. Thus we obtain a Hermitian vector bundle $(T^{1,0}\tilde{M}, \langle\ ,\ \rangle)$ over $\tilde{M}$.

The Hermitian connection $D : \Gamma\big(T^{1,0}\tilde{M}\big) \to \Gamma\big(T^*_{\mathbb{C}}\tilde{M} \otimes T^{1,0}\tilde{M}\big)$ defined uniquely on this Hermitian vector bundle $(T^{1,0}\tilde{M}, \langle\ ,\ \rangle)$ is referred to as the *Chern-Finsler connection* of F. The main feature of the Chern-Finsler connection D is its compatibility with the Hermitian metric $\langle\ ,\ \rangle$ in the sense that D preserves $\langle\ ,\ \rangle$. Namely, for all $X \in T^{1,0}\tilde{M}$ and $Y, Z \in \Gamma(T^{1,0}\tilde{M})$, the corresponding covariant differentiation ∇ satisfies

$$X\langle Y, Z\rangle = \langle \nabla_X Y, Z\rangle + \langle Y, \nabla_{\bar{X}} Z\rangle.$$

In particular, if $Z = \sum Z^\mu \dot{\partial}_\mu \in \Gamma(\mathcal{V})$ and $X = \sum X^\alpha \delta_\alpha + \sum \dot{X}^\nu \dot{\partial}_\nu \in T^{1,0}\tilde{M}$, then the covariant derivative $\nabla_X Z$ is given in local coordinates by

$$\nabla_X Z = \sum \left\{\sum X^\alpha \left[\delta_\alpha(Z^\mu) + \sum \Gamma^\mu_{\nu,\alpha} Z^\nu\right] + \sum \dot{X}^\rho \left[\dot{\partial}_\rho(Z^\mu) + \sum \Gamma^\mu_{\nu\rho} Z^\nu\right]\right\} \dot{\partial}_\mu,$$

where

$$\Gamma^\mu_{\nu\rho} = \sum G^{\bar{\tau}\mu} G_{\nu\bar{\tau}\rho} \tag{6.5}$$

and

$$\Gamma^\mu_{\nu,\alpha} = \sum G^{\bar{\tau}\mu} \delta_\alpha(G_{\nu\bar{\tau}}) = \sum G^{\bar{\tau}\mu}\left(G_{\nu\bar{\tau},\alpha} - \sum G^{\bar{\rho}\sigma} G_{\nu\bar{\tau}\sigma} G_{\bar{\rho},\alpha}\right) = \dot{\partial}_\nu(\Gamma^\mu_{,\alpha}), \tag{6.6}$$

which coincides with (5.4).

The Chern-Finsler connection D on $(T^{1,0}\tilde{M}, \langle\ ,\ \rangle)$ is not torsion-free in general. Indeed, first note that, by defining $D(\bar{Y}) = \overline{DY}$ and complex linearity, we can extend the Chern-Finsler connection D to the whole $T_C\tilde{M}$, when necessary. Then the $(2,0)$-*torsion* θ and the $(1,1)$-*torsion* τ of D are defined to be

$$\nabla_X Y - \nabla_Y X = [X,Y] + \theta(X,Y),$$
$$\nabla_X \bar{Y} - \nabla_{\bar{Y}} X = [X,\bar{Y}] + \tau(X,\bar{Y}) + \bar{\tau}(X,\bar{Y}),$$

where $X, Y \in \Gamma(T^{1,0}\tilde{M})$. The $(2,0)$-torsion θ is $\mathcal{H}$-valued and is related to the Kählerianity of D, whereas the $(1,1)$-torsion τ is $\mathcal{V}$-valued and relates to the curvature of D.

Recall that in the case of usual Hermitian metrics the vanishing of the torsion is equivalent to the metric being Kähler. However, the situation is a bit subtler for complex Finsler metrics and in fact we have three notions of Kählerianity as follows. Note that the decomposition $T^{1,0}\tilde{M} = \mathcal{V} \oplus \mathcal{H}$ induces the decomposition of the bundles of forms. Define the horizontal part $p_H\theta$ of θ as the composition of θ with the projection onto the horizontal forms, the vertical part $p_V\theta$ as that with the projection onto the vertical forms and the mixed part of θ as $\theta - p_H\theta - p_V\theta$. Then we see that $p_V\theta$ vanishes identically, the mixed part $\theta - p_H\theta$ vanishes identically if and only if $G = F^2$ is a Hermitian metric, and $\theta \equiv 0$ if and only if G is a usual Kähler metric. With these understood, we say that F is a *strongly Kähler Finsler* metric if $\theta(H_1, H_2) = 0$ for all $H_1, H_2 \in \mathcal{H}$. Also, we say that F is a *Kähler Finsler* metric if $\theta(H, \chi) = 0$ for all $H \in \mathcal{H}$, and that F is a *weakly Kähler* metric if for all $H \in \mathcal{H}$

$$\langle \theta(H,\chi), \chi \rangle = 0.$$

(Note that in Royden [21], F is called a Kähler Finsler metric if this last condition is satisfied.) In local coordinates, F is strongly Kähler if and only if $\Gamma^{\rho}_{\mu,\nu} = \Gamma^{\rho}_{\nu,\mu}$, and it is Kähler if and only if $\sum \Gamma^{\rho}_{\mu,\nu} v^{\mu} = \sum \Gamma^{\rho}_{\nu,\mu} v^{\mu}$, whereas it is weakly Kähler if and only if

$$\sum G_{\rho} \left(\Gamma^{\rho}_{\mu,\nu} - \Gamma^{\rho}_{\nu,\mu} \right) v^{\mu} = 0,$$

which is equivalent to (5.8).

The *curvature* R of the Chern-Finsler connection D is defined to be

$$R = D \circ D : \Gamma\big(T^{1,0}\tilde{M}\big) \to \Gamma\big(\Lambda^2(T^*_C\tilde{M}) \otimes T^{1,0}\tilde{M}\big).$$

Associated with R is defined the *curvature form* $\Omega \in \Gamma\big(\Lambda^2(T^*_C\tilde{M}) \otimes \Lambda^{1,0}\tilde{M} \otimes T^{1,0}\tilde{M}\big)$ of D by

$$\Omega(X,Y)Z = R(Z)(X,Y).$$

Since D is the Hermitian connection on $(T^{1,0}\tilde{M}, \langle\ ,\ \rangle)$, its curvature form Ω has neither $(2,0)$-components nor $(0,2)$-components, and satisfies the following identities:

$$\begin{aligned}
\nabla_X\nabla_Y - \nabla_Y\nabla_X &= \nabla_{[X,Y]},\\
\nabla_X\nabla_{\bar{Y}} - \nabla_{\bar{Y}}\nabla_X &= \nabla_{[X,\bar{Y}]} + \Omega(X,\bar{Y}),\\
\nabla_{\bar{X}}\nabla_{\bar{Y}} - \nabla_{\bar{Y}}\nabla_{\bar{X}} &= \nabla_{[\bar{X},\bar{Y}]},
\end{aligned}$$

where $X, Y \in \Gamma(T^{1,0}\tilde{M})$. Moreover, the $(1,1)$-torsion τ of D is related to Ω by

$$\tau(X,\bar{Y}) = \Omega(X,\bar{Y})\iota.$$

In terms of the local frame field $\{\delta_\alpha, \dot{\partial}_\mu\}$ and its dual coframe field $\{dz^\alpha, \psi^\mu\}$, where

$$\psi^\mu = dv^\mu + \sum \Gamma^\mu_{,\alpha} dz^\alpha,$$

we can write the curvature form Ω of D as [2]

$$\Omega^\mu_\nu = \sum R^\mu_{\nu,\alpha\bar{\beta}} dz^\alpha \wedge d\bar{z}^\beta + \sum R^\mu_{\nu\rho,\bar{\beta}} \psi^\rho \wedge d\bar{z}^\beta + \sum R^\mu_{\nu\bar{\rho},\alpha} dz^\alpha \wedge \overline{\psi^\rho} + \sum R^\mu_{\nu\rho\bar{\sigma}} \psi^\rho \wedge \overline{\psi^\sigma},$$

and the components of the curvature form are given by

$$\begin{aligned}
R^\mu_{\nu,\alpha\bar{\beta}} &= -\delta_{\bar{\beta}}(\Gamma^\mu_{\nu,\alpha}) - \sum \Gamma^\mu_{\nu\rho}\delta_{\bar{\beta}}(\Gamma^\rho_{,\alpha}), \quad R^\mu_{\nu\rho,\bar{\beta}} = -\delta_{\bar{\beta}}(\Gamma^\mu_{\nu\rho}) = R^\mu_{\rho\nu,\bar{\beta}},\\
R^\mu_{\nu\bar{\rho},\alpha} &= -\dot{\partial}_{\bar{\rho}}(\Gamma^\mu_{\nu,\alpha}) - \sum \Gamma^\mu_{\nu\sigma}\Gamma^\sigma_{\bar{\rho},\alpha}, \quad R^\mu_{\nu\rho\bar{\sigma}} = -\dot{\partial}_{\bar{\sigma}}(\Gamma^\mu_{\nu\rho}) = R^\mu_{\rho\nu\bar{\sigma}},
\end{aligned} \tag{6.7}$$

where $\Gamma^\mu_{\nu,\alpha}$, $\Gamma^\rho_{,\alpha}$ and $\Gamma^\mu_{\nu,\alpha}$ are defined by (6.6), (6.4) and (6.5), respectively.

7. Complex analyticity of harmonic maps

As in Section 5, let M be a complex manifold of dimension m, and suppose that M is equipped with a weakly Kähler Finsler metric $F : T^{1,0}M \to \boldsymbol{R}$. Let N be a compact Riemann surface and $f : N \to M$ a harmonic map from N to M. We denote a local coordinate system for M by $(z^1, \ldots, z^m)$ and that for N by w. Also, we denote $T^{1,0}M$ minus its zero section simply by $\tilde{M}$.

Given a smooth variation (5.3) of $f = f_0$, we now consider a map $f^\dagger : \Delta \times N \to M$ defined by $f^\dagger(t,w) = f_t(w)$ for $(t,w) \in \Delta \times N$, and set

$$T = df^\dagger\left(\frac{\partial}{\partial w}\right), \quad U = df^\dagger\left(\frac{\partial}{\partial \bar{w}}\right), \quad V = df^\dagger\left(\frac{\partial}{\partial t}\right), \quad W = df^\dagger\left(\frac{\partial}{\partial \bar{t}}\right). \tag{7.1}$$

Note that T and U are vector fields which are tangent along the variation $f^\dagger$, while V and W are transverse to the variation $f^\dagger$ in general. According to the decomposition of $T_{\boldsymbol{C}}M$ in (5.1), we may decompose these vector fields, say U, as

$$U = {}'U + {}''U = \sum \frac{\partial f^{\dagger\alpha}}{\partial \bar{w}}\frac{\partial}{\partial z^\alpha} + \sum \frac{\partial \overline{f^{\dagger\alpha}}}{\partial \bar{w}}\frac{\partial}{\partial \bar{z}^\alpha} \in T^{1,0}M \oplus T^{0,1}M.$$

[2] Note that for the components of curvature forms, our convention in this section is different, in the usage of indices, from that used in Section 2.

Similarly, we decompose T, V and W and write them as

$$T = 'T + ''T, \quad V = 'V + ''V, \quad W = 'W + ''W.$$

In order to express the second variation of the $\bar\partial$-energy $E_{\bar\partial}(f_t)$ of f_t in proper form, we need two further ingredients as in [1]. Let $D : \Gamma(T_C\tilde M) \to \Gamma(T_C^*\tilde M \otimes T_C\tilde M)$ be the Chern-Finsler connection of F, and ∇ the covariant differentiation defined by D. For the $(1,1)$-torsion τ of D, we then define the *horizontal* $(1,1)$-*torsion* $\tau^{\mathcal H}$ by

$$\tau^{\mathcal H}(X,\bar Y) = \Theta(\tau(X,\bar Y)) = \Omega(X,\bar Y)\chi,$$

where $X, Y \in \Gamma(T^{1,0}\tilde M)$. Also, on the horizontal subbundle $\mathcal H$ of $T^{1,0}\tilde M$, we define the *symmetric product* $\langle\!\langle\ ,\ \rangle\!\rangle : \mathcal H \times \mathcal H \to \boldsymbol{C}$ by

$$\langle\!\langle H_1, H_2\rangle\!\rangle_v = \sum_{\mu,\nu=1}^{m} G_{\mu\nu}(z,v)H_1^\mu H_2^\nu, \quad G_{\mu\nu} = \partial^2 G/\partial v^\mu \partial v^\nu,$$

where $H_1, H_2 \in \mathcal H_v = \mathcal H \cap \tilde\pi^{-1}(z,v)$.

With these preliminaries, the second variation formula of the $\bar\partial$-energy functional is given by the following

Theorem 4. *Let M be a complex manifold equipped with a weakly Kähler Finsler metric $F : T^{1,0}M \to \boldsymbol{R}$, and N a compact Riemann surface. Let $f : N \to M$ be a harmonic map from N to M, which is nowhere holomorphic. Let $f_t : N \to M$ be a smooth variation of $f = f_0$ parametrized by an open disk $t \in \Delta$ in $\boldsymbol{C}$. Then*

$$\begin{aligned}\frac{\partial^2}{\partial\bar t\partial t}E_{\bar\partial}(f_t)\Big|_{t=0} = & \int_N \langle\nabla_{U^H}{'V^H}, \nabla_{V^H}{'U^H}\rangle + \langle\nabla_{W^H}{'U^H}, \nabla_{U^H}{'W^H}\rangle \\ & + \int_N \langle\Omega\left('W^H, ''U^H\right)'V^H, 'U^H\rangle - \langle\Omega\left('U^H, ''W^H\right)'V^H, 'U^H\rangle \\ & + \int_N \left\langle 'U^H, \Omega(\overline{''W^H}, ''U^H)'W^H\right\rangle - \left\langle 'U^H, \Omega('U^H, \overline{'W^H})'W^H\right\rangle \\ & + \int_N \langle\!\langle\tau^{\mathcal H}\left('U^H, ''W^H\right), 'V^H\rangle\!\rangle - \langle\!\langle\tau^{\mathcal H}\left('W^H, ''U^H\right), 'V^H\rangle\!\rangle \\ & + \int_N \overline{\langle\!\langle\tau^{\mathcal H}\left('U^H, ''V^H\right), 'W^H\rangle\!\rangle} - \overline{\langle\!\langle\tau^{\mathcal H}\left('V^H, ''U^H\right), 'W^H\rangle\!\rangle},\end{aligned}$$

where the integration is performed with respect to $(\sqrt{-1}/2)dw \wedge d\bar w$, Ω is the curvature form of the Chern-Finsler connection of F, and $U^H, V^H, W^H, \ldots$ denote the horizontal lifts of $U, V, W, \ldots,$ to $T^{1,0}\tilde M$, evaluated at $t = 0$, respectively.

Concerning the complex analyticity of a $\bar\partial$-energy minimizing harmonic map, from this second variation formula we can now derive the following

Theorem 5. *Let M be a complex manifold equipped with a weakly Kähler Finsler metric F which has positive curvature. Let $f : P_1(\boldsymbol{C}) \to M$ be a $\bar\partial$-energy minimizing harmonic map from a Riemann sphere $P_1(\boldsymbol{C})$ to M. If the pull-back $f^*c_1(M)$ of*

the first Chern class of M evaluated on $P_1(\boldsymbol{C})$ is nonnegative (resp. nonpositive), then f is holomorphic (resp. antiholomorphic).

A rough sketch of the proof for the holomorphic case proceeds as follows. Let $T^{1,0}M$ be the holomorphic tangent bundle of M. Regarding $f^*T^{1,0}M$ as a holomorphic vector bundle over $P_1(\boldsymbol{C})$, we see by the result of Grothendieck [9] that $f^*T^{1,0}M$ splits into a direct sum of holomorphic line bundles L_i over $P_1(\boldsymbol{C})$. Since $f^*c_1(M)([P_1(\boldsymbol{C})]) \geq 0$ by assumption, it follows that $c_1(L_i)([P_1(\boldsymbol{C})]) \geq 0$ for some i. Hence, by the Riemann-Roch formula, we can find a nontrivial global holomorphic section $\boldsymbol{s}$ of L_i, and hence of $f^*T^{1,0}M$, over $P_1(\boldsymbol{C})$. We now construct a smooth variation f_t of $f = f_0$ such that at $t = 0$

$$'W = 0, \quad 'V = \boldsymbol{s},$$

and suppose that f is not holomorphic. It should be remarked that it follows from the equations (5.9) that if f is holomorphic on an open set of N, then it must be holomorphic everywhere. Hence we may suppose f is nowhere holomorphic, and apply Theorem 4.

Since $\boldsymbol{s}$ is holomorphic, $\nabla_{U^H}{'V^H} = 0$ at $t = 0$. Also, $\nabla_{U^H}{'W^H} = 0$ at $t = 0$. Moreover, the following identity holds for a weakly Kähler Finsler metric (cf. [1]):

$$\langle \Omega\left('U^H, ''W^H\right) 'V^H, 'U^H \rangle - \langle \Omega\left('V^H, ''W^H\right) 'U^H, 'U^H \rangle = \langle\!\langle \tau^{\mathcal{H}}('U^H, ''W^H), 'V^H \rangle\!\rangle .$$

Plugging all of these into the second variation formula in Theorem 4, we find at $t = 0$

$$\begin{aligned} \frac{\partial^2}{\partial \bar{t} \partial t} E_{\bar{\partial}}(f_t)\bigg|_{t=0} &= -\int_{P_1(\boldsymbol{C})} \langle \Omega\left('V^H, ''W^H\right)'U^H, 'U^H \rangle\rangle \\ &= -\int_{P_1(\boldsymbol{C})} \sum G_{\rho\bar{\nu}} R^{\rho}_{\mu,\alpha\bar{\beta}} \frac{\partial f^\alpha}{\partial t} \frac{\overline{\partial f^\beta}}{\partial t} \frac{\partial f^\mu}{\partial \bar{w}} \frac{\overline{\partial f^\nu}}{\partial \bar{w}}, \end{aligned} \tag{7.2}$$

where $R^{\rho}_{\mu,\alpha\bar{\beta}}$ are components of the curvature form Ω in (6.7).

Since f is $\bar{\partial}$-energy minimizing, the left-hand side of (7.2) is nonnegative. On the other hand, the positivity of the curvature of F implies that the right-hand side of (7.2) is negative, which is a contradiction. Hence we conclude that f must be holomorphic.

The detail of the proof of the first variation formula (5.6) and Theorems 4 and 5 will be published elsewhere ([20]).

References

[1] M. Abate and G. Patrizio, *Finsler Metrics – A Global Approach with applications to geometric function theory*, Lecture Notes in Math. Vol. 1591, Springer-Verlag, Berlin-Heidelberg-New York, 1994.

[2] M. Abate and G. Patrizio, Holomorphic curvature of Finsler metrics and complex geodesics, J. Geom. Anal. 3 (1996), 341–363.

[3] M. Abate and G. Patrizio, Kähler Finsler manifolds of constant holomorphic curvature, Internat. J. Math. 8 (1997), 169–186.
[4] R.L. Bishop and S.I. Goldberg, On the second cohomology group of a Kähler manifold of positive curvature, Proc. Amer. Math. Soc. 16 (1965), 119–122.
[5] P. Centore, Finsler Laplacians and minimal-energy maps, Internat. J. Math. 11 (2000), 1–13.
[6] J. Eells and L. Lemaire, *Selected Topics in Harmonic Maps*, C. B. M. S. Regional Conference Series No. 50, Amer. Math. Soc., Providence, Rhode Island, 1983.
[7] T. Frankel, Manifolds with positive curvature, Pacific J. Math. 11 (1961), 165–174.
[8] S.I. Goldberg and S. Kobayashi, Holomorphic bisectional curvature, J. Differential Geom. 1 (1967), 225–233.
[9] A. Grothendieck, Sur la classification des fibrés holomorphes sur la sphères de Riemann, Amer. J. Math. 79 (1957), 121–138.
[10] R. Hartshorne, *Ample Subvarieties of Algebraic Varieties*, Lecture Notes in Math. Vol. 156, Springer-Verlag, Berlin-Heidelberg-New York, 1970.
[11] S. Kobayashi, On compact Kähler manifolds with positive Ricci tensor, Ann. of Math. 74 (1961), 570–574.
[12] S. Kobayashi, Negativity of vector bundles and complex Finsler structures, Nagoya J. Math. 57 (1975), 153–166.
[13] S. Kobayashi, Complex Finsler vector bundles, Contemp. Math. 196 (1996), 145–152.
[14] S. Kobayashi and T. Ochiai, On complex manifolds with positive tangent bundles, J. Math. Soc. Japan 22 (1970), 499–525.
[15] S. Kobayashi and T. Ochiai, Characterizations of complex projective spaces and hyperquadrics, J. Math. Kyoto Univ. 13 (1973), 31–47.
[16] T. Mabuchi, $\boldsymbol{C}^3$-actions and algebraic three-folds with ample tangent bundle, Nagoya Math. J. 69 (1978), 33–64.
[17] M. Marcus and V. J. Mizel, Continuity of certain Nemitsky operators on Sobolev spaces and the chain rule, J. Anal. Math. 28 (1975), 303–334.
[18] X. Mo, Harmonic maps from Finsler manifolds, Illinois J. Math. 45 (2001), 1331–1345.
[19] S. Mori, Projective manifolds with ample tangent bundles, Ann. of Math. 110 (1979), 593–606.
[20] S. Nishikawa, Harmonic maps into complex Finsler manifolds, in preparation.
[21] H.L. Royden, Complex Finsler metrics, Contemp. Math. 49 (1986), 119–124.
[22] H. Rund, Generalized metrics on complex manifolds, Math. Nachr. 34 (1967), 55–77.
[23] J. Sacks and K. Uhlenbeck, The existence of minimal immersions of 2-spheres, Ann. of Math. 113 (1981), 1–24.
[24] Y.-T. Siu and S.-T. Yau, Complex Kähler manifolds of positive bisectional curvature, Invent. Math. 59 (1980), 189–204.
[25] E. Vesentini, Complex geodesics, Composito Math. 44 (1981), 375–394.

Seiki Nishikawa
Mathematical Institute, Tohoku University
980-8578 Sendai, Japan
e-mail: `nisikawa@math.tohoku.ac.jp`

Progress in Nonlinear Differential Equations
and Their Applications, Vol. 59, 133–148

Regularity of Harmonic Maps from a Flat Complex

Chikako Mese

Abstract. We show that a harmonic map from an admissible flat simplicial complex into a metric space of non-positive curvature is Lipschitz continuous away from the $(n-2)$-simplices.

1. Introduction

The study of harmonic maps into singular targets was initiated by the work of [GS] who considered locally compact polyhedral targets of non-positive curvature. In particular, these harmonic maps were shown to be Lipschitz continuous in the interior of the domain. In [KS1], the interior Lipschitz regularity result for harmonic maps was extended to the case when the target is a (not necessarily locally compact) metric space of non-positive curvature (which we will refer to as a NPC space). The interior Lipschitz regularity of harmonic maps from a Riemannian domain into metric spaces of curvature bounded from above with a smallness of image assumption was shown by [Se1].

The papers [Ch], [EF] and [F] studies the regularity of harmonic maps from singular domains. In particular, [EF] considers domain X which is an admissible Riemannian simplicial complex and a target Y which is a NPC space and proves that a harmonic map $f : X \to Y$ is pointwise Hölder continuous in the interior of the domain. The interior pointwise Hölder regularity of harmonic maps from an admissible Riemannian simplicial complex into a locally compact metric spaces of curvature bounded from above with a smallness of image assumption was shown in [F].

A natural question is to ask when a harmonic map from a singular domain is Lipschitz continuous. In this paper, we consider the interior regularity of harmonic maps from an admissible flat simplicial complex. We show

Theorem 1. *Let X be an n-dimensional admissible flat simplicial complex, Y a metric space of non-positive curvature and $f : X \to Y$ a harmonic map. Then f is Lipschitz continuous away from ∂X and the $(n-2)$-simplices of X.*

* The author is partially supported by research grant NSF DMS #0072483.

Theorem 1 shows that the Lipschitz regularity of harmonic maps can be shown for a domain that is not topologically a manifold. Although we restrict ourselves to the flat metric on the domain in this paper, the Lipschitz regularity of harmonic maps should still hold for a general admissible n-dimensional Riemannian simplicial complex with appropriate assumptions on the metric near the $(n-1)$ simplices. We conjecture that Theorem 1 can be generalized to include the case when the smooth Riemannian metric defined on each n-dimensional simplex F of X extends smoothly beyond each $(n-1)$-dimensional simplex of F. We hope to tackle this issue in a future paper. On the other hand, the result that f is Lipschitz continuous away from $(n-2)$-simplices optimal since the Lipschitz regularity property can fail at $(n-2)$-simplices. For example, the function $f(r,\theta) = r^a \cos\theta$ on a unit disk D with metric given by $g = dr^2 + \frac{1}{a^2} r^2 d\theta^2$ in polar coordinates is a harmonic map. Clearly, when $a < 1$ (which corresponds to a (D,g) being a two-dimensional metric cone with negative curvature), f is not Lipschitz at 0.

In [DM], we prove the Lipschitz regularity of harmonic maps from a two-dimensional admissible flat simplicial complex away from 0-simplices. The Lipschitz regularity is important in applications of harmonic map theory. For example, in [DM], the Lipschitz continuity was crucial in the existence and compactness theorems needed for the study of hyperbolic manifolds. The method of proof is quite different when we consider higher-dimensional domains. The argument of [DM] uses the Hopf differential and hence is strictly a two-dimensional argument.

2. Definitions

A connected locally finite n-dimensional simplicial complex is called admissible (cf. [Ch] and [EF]) if the following two conditions hold:

(i) X is dimensionally homogeneous, i.e., every simplex is contained in an n-simplex, and

(ii) X is $(n-1)$-chainable, i.e., every two n-simplices A and B can be joined by a sequence $A = F_0, e_0, F_1, e_1, \ldots, F_{k-1}, e_{k-1}, F_k = B$ where F_i is an n-simplex and e_i is an $(n-1)$-simplex contained in F_i and F_{i+1}.

The boundary ∂X of X is the union of all simplices of dimension $n-1$ which is contained in only one n-dimensional simplex. Here and henceforth, we use the convention that simplices are understood to be closed. A locally finite simplicial complex is called a Riemannian simplicial complex if a smooth bounded Riemannian metric is defined on n-simplex. This set of Riemannian metrics induces a distance function on X which we will denote by $d_X(\cdot,\cdot)$. We say a Riemannian simpicial complex is flat if:

(i) the Riemannian metric g_A on each n-simplex A makes (A, g_A) isometric to a subset of $\mathbf{R}^n$, and

(ii) if A and B are adjacent n-simplex sharing an $(n-1)$-simplex e, the metrics g_A and g_B induce the same metric g_e on e which makes (e, g_e) isometric to a subset of $\mathbf{R}^{n-1}$.

A complete metric space (Y, d) is said to have curvature bounded from above by κ if the following conditions are satisfied:

(i) The space (Y, d) is a length space. That is, for any two points P and Q in Y, there exists a rectifiable curve γ_{PQ} so that the length of γ_{PQ} is equal to $d(P,Q)$ (which we will sometimes denote by d_{PQ} for simplicity). We call such distance realizing curves geodesics.

(ii) Let $a = \sqrt{|\kappa|}$. Every point $P_0 \in Y$ has a neighborhood $U \subset Y$ so that given $P, Q, R \in U$ (assume $d_{PQ}+d_{QR}+d_{RP} < \frac{\pi}{\sqrt{\kappa}}$ for $\kappa > 0$) with Q_t defined to be the point on the geodesic γ_{QR} satisfying $d_{QQ_t} = td_{QR}$ and $d_{Q_tR} = (1-t)d_{QR}$, we have

$$\cosh(ad_{PQ_t}) \leq \frac{\sinh((1-t)ad_{QR})}{\sinh(ad_{QR})}\cosh ad_{PQ} + \frac{\sinh(tad_{QR})}{\sinh(ad_{QR})}\cosh ad_{PR}$$

for $\kappa < 0$,

$$d^2_{PQ_t} \leq (1-t)d^2_{PQ} + td^2_{PR} - t(1-t)d^2_{QR}$$

for $\kappa = 0$, and

$$\cos(ad_{PQ_t}) \geq \frac{\sin((1-t)ad_{QR})}{\sin(ad_{QR})}\cos ad_{PQ} + \frac{\sin(tad_{QR})}{\sin(ad_{QR})}\cos ad_{PR}$$

for $\kappa > 0$.

We will say that (Y, d) is NPC (non-positively curved) if it has curvature bounded from above by 0. A simply connected space of curvature bounded from above by κ is commonly referred to as a $CAT(\kappa)$ space in literature.

A map from X into Y is called harmonic if it is locally energy minimizing. Recall that, when (X^m, g) and (Y^n, h) are Riemannian manifolds, then the energy of $f : X \to Y$ is

$$E^f := \int_X |\nabla f|^2 d\mu$$

where

$$|\nabla f|^2(x) = \sum_{\alpha,\beta=1}^{m} g^{\alpha\beta}(x)h_{ij}(f(x))\frac{\partial f^i}{\partial x^\alpha}\frac{\partial f^j}{\partial x^\beta}$$

with (x^α) and (f^i) the local coordinate systems around $x \in X$ and $f(x) \in Y$ respectively.

If (X, g) is a Riemannian manifold but (Y, d) is only assumed to be a complete metric space, then we use the Korevaar-Schoen definition of energy where E^f is defined as above with $|\nabla f|^2 d\mu$ the weak limit of ϵ-approximate energy density measures. The ϵ-approximate energy density measures are measures derived from the average difference quotients. More speficially, define $e_\epsilon : X \to \mathbf{R}$ by

$$e_\epsilon(x) = \begin{cases} \int_{S(x,\epsilon)} \frac{d^2(f(x),f(y))}{\epsilon^2} \frac{d\sigma_{x,\epsilon}}{\epsilon^{n-1}} & \text{for } x \in X_\epsilon \\ 0 & \text{for } x \in X - X_\epsilon \end{cases}$$

where $\sigma_{x,\epsilon}$ is the induced measure on the ϵ-sphere $S(x,\epsilon)$ centered at x and $X_\epsilon = \{x \in X : d(x, \partial X) > \epsilon\}$. This in turn defines a family of functionals $E_\epsilon^f : C_c(X) \to \mathbf{R}$ by setting

$$E_\epsilon^f(\varphi) = \int_X \varphi e_\epsilon d\mu.$$

We say f has finite energy (or that $f \in W^{1,2}(X,Y)$) if

$$E^f := \sup_{\varphi \in C_c(X), 0 \leq \varphi \leq 1} \limsup_{\epsilon \to 0} E_\epsilon^f(\varphi) < \infty.$$

It can be shown that if f has finite energy, the measures $e_\epsilon(x)dx$ converge weakly to a measure which is absolutely continuous with respect to the Lebesgue measure. Therefore, there exists a function $e(x)$, which we call the energy density, so that $e_\epsilon(x)d\mu \rightharpoonup e(x)d\mu$. In analogy to the case of real-valued functions, we write $|\nabla f|^2(x)$ in place of $e(x)$. In particular,

$$E^f = \int_X |\nabla f|^2 d\mu.$$

For $V \in \Gamma X$ where ΓX is the set of Lipschitz vector fields on X, the directional energy measure $|f_*(V)|^2 d\mu$ is similarly defined. The real-valued L^1 function $|f_*(V)|^2$ generalizes the norm squared on the directional derivative of f.

Finally, the Korevaar-Schoen definition of energy can exteded to the case when X is an admissible Riemannian simplicial complex. Here, the energy E^f is

$$\int_X |\nabla f|^2 d\mu := \int_{\cup_{i=1}^k F_i} |\nabla f|^2 d\mu$$

where $\{F_i\}_{k=1,\dots,k}$ is the set of all top-dimensional simplices of X. The functions $|\nabla f|^2$ and $|f_*(V)|^2$ are defined for almost every point in X.

3. Summary of relevant results

The Lipschitz continuity for a harmonic map from a Riemannian domain is due to Korevaar and Schoen.

Theorem 2 ([KS1]). *Let (Ω, g) be an n-dimensional Lipschitz Riemannian domain, (Y,d) an NPC space and $f : \Omega \to Y$ a harmonic map. Then f is locally Lipschitz continuous in the interior of Ω, where the local Lipschitz constant is bounded above by*

$$C\left(\frac{E^f}{\min\{1, dist(x, \partial\Omega)\}}\right)^{\frac{1}{2}}$$

where C is a constant which depends only on n and the regularity of the metric g.

The boundary regularity result is due to Serbinowski.

Theorem 3 ([Se2]). *Let (Ω, g) be an n-dimensional Lipschitz Riemannian domain, (Y, d) an NPC space, $\phi \in W^{1,2}(\Omega, Y)$ and $f : \Omega \to Y$ a harmonic map such that $f = \phi$ on $\partial\Omega$. If, for a relatively open subset $\Gamma \subset \partial\Omega$, ϕ is locally Hölder continuous in a neighborhood of $\Gamma \cup \Omega$ with Hölder exponent α, then f is locally Hölder continuous with Hölder exponent β for every $0 \leq \beta < \alpha$.*

The Hölder regularity result is due to Eells and Fuglede.

Theorem 4 ([EF]). *Let X be an admissible Riemanian simplicial complex, (Y, d) an NPC space and $f : X \to Y$ a harmonic map. Then f is pointwise Hölder continuous; i.e., for any point $a \in X$, there exists positive constants A, α, δ depending on a so that*

$$d(f(x), f(a)) \leq A d_X(x, a)^\alpha$$

whenever $d_X(x, a) < \delta$.

4. Preliminary results

We will henceforth assume that the domain of a harmonic map is a flat 2-dimensional simplicial complex for simplicity. The proof of the Lipschitz regularity result for domain dimension equal to 2 given below generalize in a straighforward manner to the case when the domain is a flat n-dimensional simplicial complex.

Since this is a local result, we consider a harmonic map $f : X_1 \to (Y, d)$ where X_1 is isometric to a neighborhood of a point of an $(n-1)$-simplex away from $(n-2)$-simplices. We define X_1 below. Take n copies of the unit upper half disk $D^+ = \{(x, y) \in \mathbf{R}^2 | x^2 + y^2 < 1, y \geq 0\}$. We would like to distinguish these copies so we label them $D_1^+, \ldots, D_n^+$ and use (x_i, y_i) to denote the point corresponding to $(x, y) \in D^+$ on the ith copy D_i^+. Let $X_1 = \cup_{i=1}^n D_i^+ / \sim$ where $\sim$ is defined by

$$(x_i, 0) \sim (x_j, 0) \quad \text{for } x \in \mathbf{R}. \tag{1}$$

In other words, $\sim$ identifies the x-axis of D_i^+ to the x-axis of D_j^+ for all i and j. Because of this identification, we can denote $(x_i, 0)$ by $(x, 0)$. We will denote $(0, 0)$ by 0. We let $\Gamma_r = \{(x, 0) \in X_1 : -r < x < r\}$ and $\Gamma_{r,\epsilon} = \{(x_i, y_i) \in X_1 : -r < x_i < r, 0 \leq y < \epsilon\}$. Furthermore, for $z^1 = (x_i^1, y_i^1), z^2 = (x_i^2, y_i^2) \in X_1$, we let $|z^1 - z^2|$ be the distance between z^1 and z^2. In particular, $|z| = |z - 0| = \sqrt{x_i^2 + y_i^2}$. Finally, let $X_r = \{z = (x_i, y_i) \in X_1 : |z| < r\}$ and $D_{r,i}^+ = D_i^+ \cap X_r$.

We first prove that harmonic maps from X_1 is Lipschitz continuous in the direction parallel to Γ_1.

Lemma 5. *Let $f : X_1 \to Y$ be a harmonic map. For any $r < 1$, there exists a constant L dependent only on r and the energy E^f of f so that $d(f(z^1), f(z^2)) \leq L|z^1 - z^2|$ for $z^1 = (x_i^1, y_i^1), z^2 = (x_i^2, y_i^2) \in X_r$ provided that $y^1 = y^2$.*

Proof. For the vector $\frac{\partial}{\partial x}$, we will denote the directional derivative measure $|f_*(\frac{\partial}{\partial x})|^2 dxdy$ by $\left|\frac{\partial f}{\partial x}\right|^2 dxdy$ for simplicity. By [KS1],

$$\left|\frac{\partial f}{\partial x}\right|^2 (x_i, y_i) = \lim_{\epsilon \to 0} \frac{d^2(f(x_i+\epsilon, y_i), f(x_i, y_i))}{\epsilon^2}$$

for a.e. $(x_i, y_i) \in X_1$. It will be sufficient to show an upper bound for $\left|\frac{\partial f}{\partial x}\right|^2$ in X_r which is dependent only on r and E^f.

Let D_r be a disk of radius r, take n copies of D_r and label them $D_{r,1}, \ldots, D_{r,n}$. Let $\Omega_r = \cup_{i=1}^n D_{r,i}/\approx$ where the equivalence relation $\approx$ is defined so that $(x_i, 0) \approx (x_j, 0)$. For a fixed $r_0 < 1$ and $\epsilon > 0$ sufficiently small so that $r_0 + \epsilon < 1$, let $g : X_{r_0} \to Y$ be defined by $g(x_i, y_i) = f(x_i + \epsilon, y_i)$. Define $\Phi : \Omega_{r_0} \to X_{r_0}$ by setting $\Phi(x_i, y_i) = (x_i, |y_i|)$ and let $F, G : \Omega_{r_0} \to (Y, d)$ be defined by $F = f \circ \Phi$, and $G = g \circ \Phi$ respectively.

For $\eta \in C_c^\infty(D_{r_0})$, define $F_\eta, F_{1-\eta} : \Omega_{r_0} \to (Y, d)$ by setting

$$F_\eta(x_i, y_i) = (1 - \eta(x_i, y_i))F(x_i, y_i) + \eta(x_i, y_i)G(x_i, y_i)$$

and

$$F_{1-\eta}(x_i, y_i) = \eta(x_i, y_i)F(x_i, y_i) + (1 - \eta(x_i, y_i))G(x_i, y_i).$$

where $(1-t)P + tQ$ denotes the point which is fraction t of the way along the geodesic from P to Q in Y. These maps are well defined with respect to the equivalence relation $\approx$; for example,

$$\begin{aligned}
F_\eta(x_i, 0) &= (1 - \eta(x_i, 0))F(x_i, 0) + \eta(x_i, 0)G(x_i, 0) \\
&= (1 - \eta(x_i, 0))f \circ \Phi(x_i, 0) + \eta(x_i, 0)g \circ \Phi(x_i, 0) \\
&= (1 - \eta(x_i, 0))f(x,0) + \eta(x_i, 0)g(x, 0) \\
&= (1 - \eta(x_j, 0))f \circ \Phi(x_j, 0) + \eta(x_j, 0)g \circ \Phi(x_j, 0) \\
&= (1 - \eta(x_j, 0))F(x_j, 0) + \eta(x_j, 0)G(x_j, 0) \\
&= F_\eta(x_j, 0).
\end{aligned}$$

By Lemma 2.4.1 of [KS1], we see that the restriction of $F_\eta, F_{1-\eta}$ to $D_{r_0,i}$ is in $W^{1,2}$ so $F_\eta, F_{1-\eta} \in W^{1,2}(\Omega_{r_0}, Y)$ and by Lemma 2.4.2 of [KS1] (again applying the lemma to the restriction of $F_\eta, F_{1-\eta}$ to $D_{r_0,i}$), we obtain the inequality

$$\begin{aligned}
&\int_{\Omega_{r_0}} |\nabla F_\eta|^2 + \int_{\Omega_{r_0}} |\nabla F_{1-\eta}|^2 \\
&\leq \int_{\Omega_{r_0}} |\nabla F|^2 + \int_{\Omega_{r_0}} |\nabla G|^2 - 2\int_{\Omega_{r_0}} \nabla\eta \cdot \nabla d^2(F, G) + \int_{\Omega_{r_0}} Q(\eta, \nabla\eta).
\end{aligned}$$

Let $f_\eta, f_{1-\eta} : X_{r_0} \to Y$ be defined by $f_\eta = F_\eta \circ \Phi^{-1}$, $f_{1-\eta} = F_{1-\eta} \circ \Phi^{-1}$ (where we let $\Phi^{-1}(x_i, y_i) = (x_i, y_i)$) respectively. Since Φ is a local isometry almost everywhere,

$$\int_{\Omega_{r_0}} |\nabla F|^2 = 2\int_{X_{r_0}} |\nabla f|^2 \leq 2\int_{X_{r_0}} |\nabla f_\eta|^2 = \int_{\Omega_{r_0}} |\nabla F_\eta|^2$$

and

$$\int_{\Omega_{r_0}} |\nabla G|^2 = 2\int_{X_{r_0}} |\nabla g|^2 \leq 2\int_{X_{r_0}} |\nabla f_{1-\eta}|^2 = \int_{\Omega_{r_0}} |\nabla F_{1-\eta}|^2.$$

Therefore,

$$-2\int_{\Omega_{r_0}} \nabla\eta \cdot \nabla d^2(F,G) + \int_{\Omega_{r_0}} Q(\eta, \nabla\eta) \geq 0$$

and replacing η by $t\eta$, dividing by t and letting $t \to 0$, we get

$$-\int_{D_{r_0}} \nabla\eta \cdot \nabla\delta \geq 0$$

for any $\eta \in C_c^\infty(D_{r_0})$ and where

$$\delta(x,y) = \sum_{i=1}^n d^2(F(x_i,y_i), G(x_i,y_i)).$$

By the mean value inequality for subharmonic functions,

$$\delta(x,y) \leq \frac{1}{\pi\rho^2}\int_{D_\rho(x,y)} \delta$$

for any $\rho \leq r_0 - \sqrt{x^2+y^2}$ where $D_\rho(x,y)$ is the disk of radius ρ centered at (x,y).

Now suppose $r < r_0$. Let $(x_i, y_i) \in X_r$ and $\rho = r_0 - r$. (As before, D_{r_0} is the disk of radius r_0 centered at the origin.) Then

$$\begin{aligned} d^2(f(x_i,y_i), g(x_i,y_i)) &= d^2(F(x_i,y_i), G(x_i,y_i)) \\ &\leq \delta(x,y) \\ &\leq \frac{1}{\pi\rho^2}\int_{D_\rho(x,y)} \delta \\ &= \frac{1}{\pi\rho^2}\int_{D_{r_0}} \delta \\ &= \frac{1}{\pi\rho^2}\int_{\Omega_{r_0}} d^2(F,G) \\ &= \frac{2}{\pi\rho^2}\int_{X_{r_0}} d^2(f,g). \end{aligned}$$

Thus, for $(\bar{x}_i, \bar{y}_i) \in X_r$ and $\rho = \frac{1-r}{2}$ and $r_0 = r + \rho$.

$$
\begin{aligned}
\left|\frac{\partial f}{\partial x}\right|^2(\bar{x}_i,\bar{y}_i) &= \lim_{\epsilon\to 0}\frac{d^2(f(\bar{x}_i,\bar{y}_i),f(\bar{x}_i+\epsilon,\bar{y}_i))}{\epsilon^2}\\
&\le \lim_{\epsilon\to 0}\sum_{j=1}^{n}\frac{d^2(f(\bar{x}_j,\bar{y}_j),f(\bar{x}_j+\epsilon,\bar{y}_j))}{\epsilon^2}\\
&\le \frac{2}{\pi\rho^2}\sum_{j=1}^{n}\lim_{\epsilon\to 0}\frac{1}{\epsilon^2}\int_{X_{r_0}} d^2(f(x,y),f^j(x+\epsilon,y))dxdy\\
&\le \frac{2}{\pi\rho^2}\int_{X_{r_0}}\left|\frac{\partial f}{\partial x}\right|^2 dxdy\\
&\le \frac{2}{\pi\rho^2}E^f. \qquad \square
\end{aligned}
$$

Using Lemma 5, we now show that a harmonic map $f : X_1 \to (Y,d)$ is β-Hölder continuous in X_r for any $\beta < 1$ and $r < 1$.

Lemma 6. *Let $f : X_1 \to Y$ be a harmonic map. For any $\beta < 1$ and $r < 1$, there exists a constant c_β so that $d(f(z_1), f(z_2)) \le c_\beta |z^1 - z^2|^\beta$ for $z^1, z^2 \in X_r$.*

Proof. By Lemma 5, f restricted to Γ_r is Lipschitz continuous. Thus, for each i, we can construct a $W^{1,2}$ map $\phi : D_i^+ \to Y$ so that $f = \phi$ on ∂D_i^+ and ϕ is Lipschitz continuous in a neighborhood of Γ_r as the following shows. Let D be the unit disk and $\psi : D \to D_i^+$ a bi-Lipschitz map. With (r,θ) the polar coordinates of D, define $\Phi : D \to Y$ by setting $\Phi(r,\theta)$ to be the point on the geodesic from $f(0)$ to $f \circ \psi(1,\theta)$ that is a distance r from $f(0)$. The NPC condition on Y guarantees that Φ is Lipschitz continuous near $\psi^{-1}(\Gamma_r)$ and hence $\phi = \Phi \circ \psi^{-1}$ is also Lipschitz near Γ_r and by construction $\phi = f$ on ∂D_i^+. Now Lemma 6 follows from Theorem 3. $\square$

5. The monotonicity formula

The following monotonicity property will be crucial in our regularity proof. For $-1 < x < 1$ and $0 < r < 1 - x$, let $B_r(x,0)$ be a ball of radius r at $(x,0) \in X_1$ and set

$$E_x^f(r) = \int_{B_r(x,0)} |\nabla f|^2 d\mu$$

and

$$I_x^f(r) = \int_{\partial B_r(x,0)} d^2(f, f(x,0))ds.$$

Lemma 7. *Let $f : X_1 \to Y$ be a harmonic map. Then*

$$r \mapsto Ord(x,r) = \frac{rE_x^f(r)}{I_x^f(r)}$$

is a non-decreasing function for $0 \le r \le 1 - x$.

Proof. For simplicity, we assume $x = 0$ and set $E(r) = E_0^f(r)$ and $I(r) = I_0^f(r)$. Let $\eta : X \to \mathbf{R}_+ \cup \{0\}$ be a continuous function which is smooth on each face of X_1. For η with $spt(\eta) \subset X_r$ and t sufficiently small, we define $F_t : X \to X$ as

$$F_t(x_i, y_i) = ((1 + t\eta(x_i, y_i))x_i, (1 + t\eta(x_i, y_i))y_i).$$

With that, we can now follow the usual calculation to prove Lemma 7. In other words, the standard computation (see [GS], Section 2 for example) done on each face of X_1 gives

$$E'(r) = 2\int_{\partial X_r} \left|\frac{\partial f}{\partial r}\right|^2 ds \tag{2}$$

for a.e. $0 \leq r \leq 1$. Again, for a.e. $0 \leq r \leq 1$, standard computation on each face of X gives,

$$I'(r) = \int_{\partial X_r} \frac{\partial}{\partial r} d^2(f, f(p_0))ds + \frac{I(r)}{r}.$$

Using the inequality $|\frac{\partial}{\partial r} d(f, f(p_0))| \leq |\frac{\partial f}{\partial r}|$ and the Schwarz inequality, the above two equations imply that

$$\frac{d}{dr}\log\left(\frac{rE(r)}{I(r)}\right) = \frac{1}{r} + \frac{E'(r)}{E(r)} - \frac{I'(r)}{I(r)} \geq 0$$

for a.e. $0 \leq r \leq 1$. □

Lemma 8. *Let $f : X_1 \to (Y, d)$ be a harmonic map so that*

$$\alpha = \lim_{\sigma \to 0} \frac{\sigma E_0^f}{I_0^f(\sigma)}.$$

Then

$$\frac{2\alpha + 1}{\sigma} \leq \frac{\frac{dI_0^f}{ds}(\sigma)}{I_0^f(\sigma)} \tag{3}$$

and hence

$$\frac{d}{d\sigma}\left(\frac{I_0^f(\sigma)}{\sigma^{2\alpha+1}}\right) \geq 0.$$

Proof. The argument of [GS], Section 2 (see also [EF]) implies

$$2\int_{X_1} |\nabla f|^2 \eta dxdy \leq -\int_{X_1} \nabla d^2(f, f(0)) \cdot \nabla \eta dxdy \tag{4}$$

for any $W^{1,2}$ function η with compact support in X_1. Choosing η to approximate the characteristic function of X_r,

$$2E_0^f(\sigma) \leq \int_{\partial B_\sigma(p_0)} \frac{\partial}{\partial r} d^2(f, f(v))ds = \frac{d}{d\sigma} I_0^f(\sigma) - \frac{1}{\sigma} I_0^f(\sigma).$$

The monotonicity property of $\sigma \mapsto Ord(0, \sigma)$ implies

$$\frac{\alpha I(\sigma)}{\sigma} \leq E(\sigma)$$

and by combining the two inequalities, we get

$$2\alpha I_0^f(\sigma) \leq \sigma \frac{d}{d\sigma} I_0^f(\sigma) - I_0^f(\sigma)$$

which lead to the desired inequality. □

We prove the following regularity result based on the order α of a harmonic map.

Theorem 9. *Let $f : X_1 \to (Y, d)$ be a harmonic map and*

$$\alpha = \lim_{\sigma \to 0} \frac{\sigma E_0^f(\sigma)}{I_0^f(\sigma)}.$$

then f satisfies

$$f(z) \leq C|z|^\alpha$$

for all $z = (x_i, y_i) \in X_{\frac{1}{2}}$ where C depends only on α and E^f.

Proof. For simplicity, we let $E(\sigma) = E_0^f(\sigma)$ and $I(\sigma) = I_0^f(\sigma)$. From Lemma 8, we have

$$\frac{2\alpha + 1}{\sigma} \leq \frac{I'(\sigma)}{I(\sigma)}$$

and integrating this differential inequality from $\sigma \in (0,1)$ to 1, we obtain

$$\frac{I(\sigma)}{\sigma} \leq \sigma^{2\alpha} I(1).$$

The monotonicity property of $\sigma \mapsto Ord(0, \sigma)$ implies

$$I(1) \leq \frac{E(1)}{\alpha},$$

and hence

$$\frac{I(\sigma)}{\sigma} \leq \frac{E(1)}{\alpha} \sigma^{2\alpha}.$$

Define $\delta : D \to \mathbf{R}$ be setting

$$\delta(x, y) = \begin{cases} \sum_{i=1}^N d^2(f(x^i, y^i), 0) & y_i \geq 0 \\ \sum_{i=1}^N d^2(f(x^i, |y^i|), 0) & y_i < 0. \end{cases}$$

Using the argument of the proof of Lemma 5 with $g(x_i, y_i) = 0$ instead of $g(x_i, y_i) = f(x_i + \epsilon, y_i)$, we can show that δ is a subharmonic function and the mean value inequality implies

$$\sup_{(x_i, y_i) \in X_{\frac{\sigma}{2}}} d^2(f(x_i, y_i), f(0)) \leq \sup_{z \in D_{\frac{\sigma}{2}}} \delta(x, y) \leq \frac{4}{\pi\sigma} \int_{\partial D_\sigma} \delta ds \leq \frac{4I(\sigma)}{\pi\sigma} \leq C\sigma^{2\alpha}$$

where $C = \frac{4E(1)}{\alpha\pi}$. □

Therefore, the Lipschitz continuity of $f : X_1 \to Y$ if we show that $\alpha \geq 1$. Hence, we will now assume that $\alpha < 1$ and show that this leads to a contradiction.

6. The tangent map

Lemma 10. *Let $f : X_1 \to Y$ be a harmonic map. Fix $0 < \beta < 1$. For any $x \in [-\frac{1}{2}, \frac{1}{2}]$ and $\sigma < \frac{1}{4}$, there exists a constant c_1 so that $E_x^f(\sigma) \le c_1\sigma^{2\beta}$.*

Proof. Let

$$k_1 = \max_{x\in[-\frac{1}{2},\frac{1}{2}]} \frac{E_x(\frac{1}{4})}{4I_x(\frac{1}{4})}.$$

By Lemma 6, there exists k_2 so that $d(f(z^1), f(z^2)) \le k_2|z^1 - z^2|^\beta$ for z^1, z^2 in a neighborhood of $\Gamma_{\frac{3}{4}}$. Thus, $I_x(\sigma) \le k_2\sigma^{2\beta+1}$ for $x \in [-\frac{1}{2}, \frac{1}{2}]$ and $\sigma < \frac{1}{4}$. The monotonicity property of Lemma 7 shows that

$$E_x(\sigma) \le \frac{E_x(\frac{1}{4})}{4I_x(\frac{1}{4})} \frac{I_x(\sigma)}{\sigma} = c_1\sigma^{2\beta}$$

where $c_1 = k_1k_2$. □

An important consequence of the monotonicity property of Lemma 7 is the existence of a tangent map.

Lemma 11. *Let $f : X_1 \to (Y, d)$ be a harmonic map,*

$$\mu^f(\lambda) = \sqrt{\frac{\lambda}{I_0^f(\lambda)}}$$

and $d_\lambda(\cdot,\cdot) = \mu^f(\lambda)d(\cdot,\cdot)$. Define $f_\lambda : X_1 \to (Y, d_\lambda)$ by setting

$$f_\lambda(x_i, y_i) = f(\lambda x_i, \lambda y_i).$$

Assume that

$$\alpha = \lim_{\sigma\to 0} \frac{\sigma E_0^f(\sigma)}{I_0^f(\sigma)} < 1.$$

There exists $\lambda_k \to 0$, an NPC space (Y_*, d_*) and a harmonic map $f_* : X_1 \to (Y_*, d_*)$ so that $d_{\lambda_k}(f_{\lambda_k}(\cdot), f_{\lambda_k}(\cdot))$ converges uniformly to $d_*(f_*(\cdot), f_*(\cdot))$ and $E^{f_{\lambda_k}}(r) \to E^{f_*}(r)$ for $0 < r \le 1$.

Proof. Let $\lambda_k \to 0$. Since

$$E_0^{f_{\lambda_k}}(1) = (\mu^f(\lambda_k))^2 E_0^f(\lambda_k) = \frac{\lambda_k E_0^f(\lambda_k)}{I_0(\lambda_k)} \le \frac{E_0^f(1)}{I_0^f(1)},$$

the energy of f_{λ_k} is uniformly bounded. The local Hölder constant and exponent of a harmonic map from X_1 to Y can be shown to be dependent only on the energy of the map and the distance to ∂X_1. (For example, we can use the proof of [Ch]. We note that although the proof of [Ch] does not apply to a general metric as is considered in [EF], it works in our case when the domain is flat. In [Ch], we clearly see the dependence of the Hölder constant and exponent only on the total energy and the distance to the boundary.) Therefore, f_{λ_k} satisfies the uniform modulus of continuity control needed to apply Proposition 3.7 of [KS2].

For $\lambda_k \to 0$ and any $\tau < 1$, we see that there exists a subsequence of f_{λ_k} which converges locally uniformly in the pullback sense to a harmonic map by applying the argument of [KS2] Proposition 3.7. Pick a sequence $\tau_n \to 1$ and by a diagonalization procedure, we can pick a subsequence f_{λ_k} (which we again denote by $\{f_{\lambda_k}\}$ by an abuse of notation) which converges locally uniformly in the pull back sense to a harmonic map $f_* : X_1 \to (Y, d_*)$. In other words, $d_{\lambda_k}(f_{\lambda_k}(\cdot), f_{\lambda_k}(\cdot))$ converges uniformly to $d_*(f_*(\cdot), f_*(\cdot))$.

We first show that f_* is non-constant. Let $E^k(r) = E_0^{f_{\lambda_k}}(r)$, $I^k(r) = I_0^{f_{\lambda_k}}(r)$, $E^*(r) = E_0^{f_*}(r)$, $I^*(r) = I_0^{f_*}(r)$ for simplicity. Repeating the computation of [GS], proof of Proposition 3.3,

$$I^k(r_0) - I^k(\theta) \leq \epsilon E^k(r_0) + \left(\frac{1}{\epsilon} + \frac{1}{\theta}\right) \int_\theta^{r_0} I^k(r)dr$$

for $0 < \theta < r < 1$ and any $\epsilon > 0$. By Lemma 7,

$$\frac{r_0 E^k(r_0)}{I^k(r_0)} \leq \frac{E^k(1)}{I^k(1)} = C$$

and hence

$$I^k(r_0) - I^k(\theta) \leq \frac{\epsilon C I^k(r_0)}{r_0} + \left(\frac{1}{\epsilon} + \frac{1}{\theta}\right) \int_\theta^{r_0} I^k(r)dr.$$

For any $\theta \in [\frac{1}{2}, 1)$, pick $r_0 \in (\theta, 1]$. Then $r_0 \geq \theta \geq \frac{1}{2}$ and by choosing $\epsilon = \frac{1}{4C}$,

$$\begin{aligned} \frac{1}{2} I^k(r_0) - I^k(\theta) &\leq (4C + \frac{1}{\theta}) \int_\theta^{r_0} I^k(r)dr \\ &\leq (4C + 2) \int_\theta^1 I^k(r)dr \\ &\leq (4C + 2)(1 - \theta) \sup_{r \in [\theta, \sigma]} I^k(r). \end{aligned}$$

Since r_0 is an arbitrary point in $(\theta, 1]$, we have

$$\frac{1}{2} \sup_{r \in [\theta, 1]} I^k(r) - I^k(\theta) \leq (4C + 2)(1 - \theta) \sup_{r \in [\theta, \sigma]} I^k(r).$$

Now choose θ sufficiently close to 1 so that $(4C + 2)(1 - \theta) \leq \frac{1}{6}$. Then

$$I^k(\theta) \geq \frac{1}{3} \sup_{r \in [\theta, 1]} I^k(r) \geq \frac{1}{3} I^k(1) = \frac{1}{3}.$$

By the uniform convergence of $d_{\lambda_k}(f_{\lambda_k}(\cdot), f_{\lambda_k}(\cdot))$ to $d_*(f_*(\cdot), f_*(\cdot))$, we then have

$$I^*(\theta) \geq \frac{1}{3}$$

and this shows that f_* is non-constant.

Finally, we show that $E^k(r) \to E^*(r)$ for $r < 1$. By Theorem 3.11 of [KS2], it will be enough to show that

$$E^k_{\epsilon-\text{strip}} := \int_{\Gamma_{1,\epsilon}} |\nabla f_k|^2 dxdy < C_1 \epsilon^\gamma$$

for $\gamma > 0$. Choose β so that $\max\{\alpha, \frac{1}{2}\} < \beta < 1$. Let $\epsilon > 0$ be given. For $\lambda < \frac{1}{4}$, let

$$E^f_{\epsilon\lambda-\text{strip}} = \int_{\Gamma_{\lambda,\lambda\epsilon}} |\nabla f|^2 dxdy$$

Note that $\Gamma_{\lambda,\lambda\epsilon}$ can be covered by $\frac{10}{\epsilon}$ number of balls of radius $2\lambda\epsilon$ centered at points on $\Gamma_\lambda \subset \Gamma_{\frac{1}{2}}$. Therefore, by Lemma 10

$$E^f_{\epsilon\lambda-\text{strip}} \leq \frac{10}{\epsilon} c_1 (2\lambda\epsilon)^{2\beta}.$$

Since

$$\alpha = \lim_{\sigma \to 0} \frac{\sigma E_0^f(\sigma)}{I_0^f(\sigma)}.$$

we know

$$\alpha I_0^f(\sigma) \leq \sigma E_0^f(\sigma)$$

by Lemma 7.

Now letting $C_2 = \lim_{\sigma\to 0} \frac{I_0^f(\sigma)}{\sigma^{2\alpha+1}}$, we see that $C_2 \leq \frac{I_0^f(\lambda_k)}{\lambda_k^{2\alpha+1}}$ by Lemma 8 or equivalently

$$(\mu^f(\lambda_k))^2 = \frac{\lambda_k}{I_0^f(\lambda_k)} \leq C_2 \lambda_k^{-2\alpha}.$$

By the definition of $f_{\lambda_k} : X_1 \to (Y, d_{\lambda_k})$, we have

$$E^k_{\epsilon-\text{strip}} = (\mu^f(\lambda_k))^2 E^f_{\epsilon\lambda_k-\text{strip}} \leq C_2 \lambda_k^{-2\alpha} \frac{10}{\epsilon} c_1 (2\lambda_k \epsilon)^{2\beta} = 10 C_2 c_1 \lambda_k^{2\beta-2\alpha} \epsilon^{2\beta-1}$$

We have chosen β so that $2\beta - 2\alpha > 0$ and $2\beta - 1 > 0$, so letting $C_1 = 10C_2c_1$ and $\gamma = 2\beta - 1$, we have

$$E^k_{\epsilon-\text{strip}} = k_3 \epsilon^\gamma.$$

By Theorem 3.11 of [KS1], this shows that the energy of f_k on each $D_i^+ \cap X_r$ converges to that of f_* so $E^k(r) \to E^{f_*}(r)$ for all $r < 1$ and hence $E^{f_{\lambda_k}}(1) \to E^{f_*}(1)$ also. □

We call f_* of Lemma 11 a tangent map of f. We note the following property of f_*.

Lemma 12. *Let $f : X_1 \to (Y, d)$ be a harmonic map so that*

$$\alpha = \lim_{\sigma \to 0} \frac{\sigma E_0^f(\sigma)}{I_0^f(\sigma)} < 1.$$

Then its tangent map $f_* : X_1 \to (Y, d_*)$ is a homogeneous map of order α; in other words,

$$d_*(f_*(z), f_*(0)) = |z|^\alpha d_* \left(f_* \left(\frac{z}{|z|} \right), f_*(0) \right).$$

Proof. Since

$$E_0^{f_{\lambda_k}}(r) = (\mu^f(\lambda_k))^2 E_0^f(\lambda_k r) = \frac{\lambda_k}{I_0(\lambda_k)} E_0^f(\lambda_k r)$$

and

$$I_0^{f_{\lambda_k}}(r) = (\mu^f(\lambda_k))^2 \frac{I_0^f(\lambda_k r)}{\lambda_k} = \frac{\lambda_k}{I_0^f(\lambda_k)} \frac{I_0^f(\lambda_k r)}{\lambda_k} = \frac{I_0^f(\lambda_k)}{I_0^f(\lambda_k r)},$$

we have

$$\frac{r E_0^{f_*}(r)}{I_0^{f_*}(r)} = \lim_{k \to \infty} \frac{r E_0^{f_{\lambda_k}}(r)}{I_0^{f_{\lambda_k}}(r)} = \lim_{k \to \infty} \frac{\lambda_k r E_0^f(\lambda_k r)}{I_0^f(\lambda_k r)} = \alpha.$$

Therefore, by proof of Lemma 3.2 of [GS], f_* is a homogeneous map of order α. □

7. The Lipschitz continuity

We are now ready to prove:

Lemma 13. *The harmonic map $f : X_1 \to Y$ is Lipschitz continuous at 0 with the Lipschitz constant dependent only on E^f.*

Proof. By Theorem 9, it is sufficient to show that $\alpha \geq 1$. Suppose $\alpha < 1$. Let $f_* : X_1 \to Y_*$ be a tangent map of f. By Lemma 5, f_* satisfies

$$d_*^2(f_*(x_i, y_i), f_*(0, y_i)) \leq L^2 x_i^2.$$

Let $f_{**} : X_1 \to Y_{**}$ be a tangent map of f_* so that for $\lambda_k \to 0$, $f_{*,\lambda_k} : X_1 \to (Y_*, d_{*,\lambda_k})$ converges in the sense of Lemma 11 to f_{**}. By Lemma 12, both f_* and f_** are homogeneous maps of order α. Then

$$\begin{aligned} & d_{**}(f_{**}(x_i, y_i), f_{**}(0, y_i)) \\ = & \lim_{k \to \infty} d_{*,\lambda_k}(f_{*,\lambda_k}(x_i, y_i), f_{*,\lambda_k}(0, y_i)) \\ = & \lim_{k \to \infty} \mu^{f_*}(\lambda_k) d_*(f_*(\lambda_k x_i, \lambda_k y_i), f_*(0, \lambda_k y_i)) \\ = & \lim_{k \to \infty} \mu^{f_*}(\lambda_k) L \lambda_k x_i \end{aligned}$$

Furthermore,

$$I^{f_*}(\lambda_k) = \int_{\partial X_{\lambda_k}} d^2(f_*, f_*(0)) ds = \lambda_k^{2\alpha+1} \int_{\partial X_1} d_*^2(f_*, f_*(0)) ds = \lambda_k^{2\alpha+1} I^{f_*}(1)$$

so we have

$$\begin{aligned}\mu^{f_*}(\lambda_k) &= \sqrt{\frac{\lambda_k}{I^{f_*}(\lambda_k)}} \\ &= \lambda_k^{-\alpha}(I^{f_*}(1))^{-\frac{1}{2}}.\end{aligned}$$

Therefore,

$$d_{**}(f_{**}(x_i, y_i), f_{**}(0, y_i)) = \lim_{k\to\infty} L\lambda_k^{1-\alpha} x_i (I^{f_*}(1))^{-\frac{1}{2}} = 0.$$

This, in particular, shows that $t \mapsto f_{**}(t_i, y_i)$ is a geodesic curve. Therefore, $\alpha = 1$ and this is a contradiction. Therefore, $\alpha \geq 1$. It follows from the proof of Theorem 9 that the Lipschitz constant depends only on E^f. □

Furthermore, it is now straightforward to show that the Lipschitz constant of a harmonic map $f : X \to Y$ at point $(x, 0)$ for $|x| < \frac{1}{2}$ depends only on E^f. We now show:

Lemma 14. *The local Lipschitz constant of $f : X_1 \to Y$ at a point $z_0 = (x_i, y_i) \in X_{\frac{1}{2}}$ with $y_i \neq 0$ depends only on E^f.*

Proof. Let $r_0 = y_i$ and

$$E^f_{(x_i,y_i)}(r) = \int_{B_r(x_i,y_i)} |\nabla f|^2 d\mu$$

and

$$I^f_{(x_i,y_i)}(r) = \int_{\partial B_r(x_i,_i)} d^2(f, f(x_i, y_i))ds.$$

Then

$$r \mapsto \frac{rE^f_{(x_i,y_i)}(r)}{I^f_{(x_i,y_i)}(r)}$$

is monotone for $r < r_0$ since $B_r(x_i, y_i)$ is contained in a 2-simplex for $r < r_0$. As in Lemma **??**, we can deduce that

$$\frac{I^f_{(x_i,y_i)}(r)}{r} \leq \frac{I^f_{(x_i,y_i)}(r_0)}{r_0^{2\alpha+1}} r^{2\alpha} \leq \frac{E^f_{(x_i,y_i)}(r_0)}{\alpha r_0^{2\alpha}} r^{2\alpha}$$

where $\alpha = \alpha(x_i, y_i) = \lim_{\sigma\to 0} \frac{\sigma E^f_{(x_i,y_i)}(\sigma)}{I^f_{(x_i,y_i)}(\sigma)}$. Therefore,

$$\begin{aligned}\sup_{z\in B_{\frac{r}{2}}(z_0)} d^2(f(z), f(z_0)) &\leq \frac{4I^f_{(x_i,y_i)}(r)}{\pi r} \\ &\leq \frac{4E^f_{(x_i,y_i)}(r_0)}{\pi\alpha r_0^{2\alpha}} r^{2\alpha}\end{aligned}$$

We know that $\alpha \geq 1$ since $(x_i, y_i) \mapsto \alpha(x_i, y_i)$ is upper semicontinuous since it is a decreasing limit of continuous functions. If $\alpha > 1$, then

$$0 \leq \lim_{z \mapsto z_0} \frac{d^2(f(z), f(z_0))}{|z - z_0|^2} \leq \frac{4E^f_{(x_i,y_i)}(r_0)}{\pi \alpha r_0^{2\alpha}} \lim_{r \to 0} r^{2\alpha - 2} = 0$$

If $\alpha = 1$, then

$$\lim_{z \mapsto z_0} \frac{d^2(f(z), f(z_0))}{|z - z_0|^2} \leq \frac{4E^f_{(x_i,y_i)}(r_0)}{\pi r_0^2} \leq \frac{16E^f_{(x_i,0)}(2r_0)}{\pi(2r_0)^2} \leq \frac{16E^f}{\pi}$$

This shows that the energy density function $|\nabla f|^2$ is uniformly bounded by a constant dependent on energy and proves Lemma 14. □

Lemma 13 and Lemma 14 combines to prove Theorem 1 for two-dimensional domain. As mentioned previously, it is straightforward to generalize the arguments here to a higher-dimensional domain.

References

[Ch] J. Chen. *On energy minimizing mappings between and into singular spaces.* Duke Math. J. 79 (1995), 77–99.

[DM] G. Daskalopoulos and C. Mese. *Harmonic maps from 2-complexes and geometric actions on trees.* Preprint.

[EF] J. Eells and B. Fuglede. *Harmonic maps between Riemannian polyhedra.* Cambridge Tracts in Mathematics 142, Cambridge University Press, Cambridge 2001.

[F] B. Fuglede. *Hölder continuity of harmonic maps from Riemannian polyhedra to spaces of upper bounded curvature.* Preprint.

[GS] M. Gromov and R. Schoen. *Harmonic maps into singular spaces and p-adic superrigidity for lattices in groups of rank one.* IHES Publ. Math. 76 (1992) 165–246.

[KS1] N. Korevaar and R. Schoen. *Sobolev spaces and harmonic maps for metric space targets.* Communications in Analysis and Geometry 1 (1993), 561–659.

[KS2] N. Korevaar and R. Schoen. *Global existence theorem for harmonic maps to non-locally compact spaces.* Communications in Analysis and Geometry 5 (1997), 333–387.

[Se1] T. Serbinowski. *Harmonic maps into metric spaces with curvature bounded above.* Thesis, University of Utah, 1995.

[Se2] T. Serbinowski. *Boundary regularity of harmonic maps to nonpositively curved metric spaces.* Comm. Anal. Geom. 2 (1994) 139–154.

Chikako Mese
Department of Mathematics
Connecticut College
New London, CT 06320, USA
fax: 860-439-2700
e-mail: cmes@conncoll.edu

For orders originating from all over the world except USA/Canada/Latin America:

Birkhäuser Verlag AG
c/o Springer GmbH & Co
Haberstrasse 7
D-69126 Heidelberg
Fax: +49 / 6221 / 345 4 229
e-mail: birkhauser@springer.de
http://www.birkhauser.ch

For orders originating in the USA/Canada/Latin America:

Birkhäuser
333 Meadowland Parkway
USA-Secaucus
NJ 07094-2491
Fax: +1 / 201 / 348 4505
e-mail: orders@birkhauser.com
http://www.birkhauser.com

For orders originating from all over the world except USA/Canada/Latin America:

Birkhäuser Verlag AG
c/o Springer GmbH & Co
Haberstrasse 7
D-69126 Heidelberg
Fax: +49 / 6221 / 345 4 229
e-mail: birkhauser@springer.de
http://www.birkhauser.ch

For orders originating in the USA/Canada/Latin America:

Birkhäuser
333 Meadowland Parkway
USA-Secaucus
NJ 07094-2491
Fax: +1 / 201 / 348 4505
e-mail: orders@birkhauser.com
http://www.birkhauser.com

MIX
Papier aus verantwortungsvollen Quellen
Paper from responsible sources
FSC® C105338

If you have any concerns about our products,
you can contact us on
ProductSafety@springernature.com

In case Publisher is established outside the EU,
the EU authorized representative is:
Springer Nature Customer Service Center GmbH
Europaplatz 3, 69115 Heidelberg, Germany

Printed by Libri Plureos GmbH
in Hamburg, Germany